β Beta

Focus: Multiple-Digit Addition and Subtraction

Student Text

By Miriam Homer

Math·U·See

1-888-854-MATH (6284)
www.MathUSee.com

Math·U·See

1-888-854-MATH (6284)
www.MathUSee.com

Copyright © 2009 by Steven P. Demme

Graphic Design by Christine Minnich
Illustrations by Gregory Snader

Printed in the United States of America

Beta β

	LESSON PRACTICE			SYSTEMATIC REVIEW			
	A	B	C	D	E	F	TEST
1 Place Value							
2 Sequencing							
3 Inequalities							
4 Round to 10							
5 Add Multiple Digit							
6 Skip Count 2							
7 Regrouping							
Unit Test 1							
8 Skip Count 10, Coins							
9 Skip Count 5, Nickel							
10 Money							
11 Round to 100							
12 Add Money							
13 Column Addition							
14 Measure: Foot							
15 Perimeter							
Unit Test 2							
16 1,000s							
17 Round to 1,000s							
18 Mult. Digit Column 1							
19 Mult. Digit Column 2							
20 Mult. Digit Subtract							
21 Time: Minutes							
22 Regrouping							
Unit Test 3							
23 Time: Hours							
24 Subtract 3 Digits							
25 Ordinal Numbers							
26 Subtract 4 Digits							
27 Subtract Money							
28 Subtract Mult. Digit							
29 Read Gauges							
30 Graphs							
Unit Test 4							
Final Test							

Count and write the number, and then say it.

1.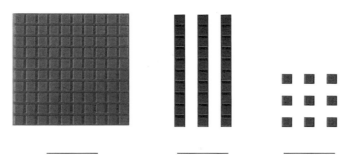

   _____   _____   _____

2.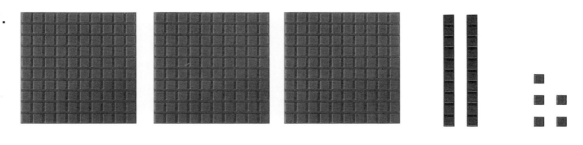

   _____   _____   _____

Build and say the number.

3. 31   4. 228

Add. These problems review adding 8 and 9.

5.   9
    + 3

6.   8
    + 7

7.   9
    + 5

8.   8
    + 3

9.   5
    + 8

10.   6
    + 9

11.   9
    + 4

12.   8
    + 9

13. 9 + 7 = _____

14. 6 + 8 = _____

15. 8 + 4 = _____

1C

Count and write the number, and then say it.

1.
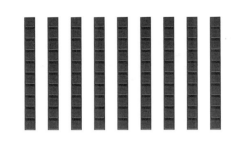

_____   _____   _____

2.

_____   _____   _____

Build and say the number.

3. 49

4. 283

Add. These problems review doubles and doubles plus one.

5.   8
     + 8

6.   7
     + 7

7.   7
     + 6

8.   4
     + 4

9.   5
     + 5

10.   6
     + 5

11.   3
     + 3

12.   2
     + 2

13. 6 + 6 = _____

14. 3 + 4 = _____

15. 9 + 9 = _____

Count and write the number, and then say it.

1.

      _____   _____   _____

2.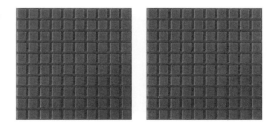

      _____   _____   _____

Build and say the number.

3. 72   4. 417

Add. These problems review making 9 and making 10.

5. 8
   + 1
   ———

6. 7
   + 3
   ———

7. 3
   + 6
   ———

8. 1
   + 9
   ———

9. 6
   + 4
   ———

10. 5
    + 5
    ———

11. 7
    + 2
    ———

12. 5
    + 4
    ———

13. 2 + 8 = _____

14. 3 + 7 = _____

15. 4 + 6 = _____

1E

Count and write the number, and then say it.

1.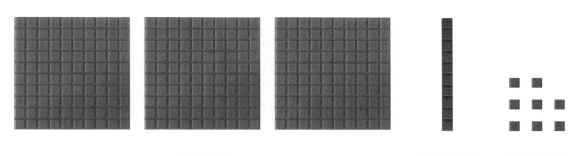

   _____   _____ _____

2.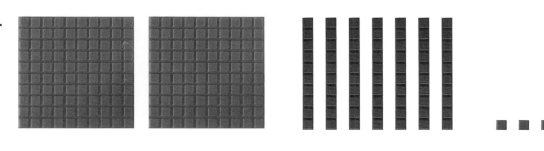

   _____   _____   _____

Build and say the number.

3. 50

4. 190

Add. These problems review 4 + 7, 5 + 7, 3 + 5, and other facts.

5.　　　4
　　　+ 7

6.　　　7
　　　+ 5

7.　　　3
　　　+ 5

8.　　　5
　　　+ 7

9.　　　5
　　　+ 3

10.　　　7
　　　+ 4

11.　　　9
　　　+ 8

12.　　　5
　　　+ 3

13.　7 + 6 = _____

14.　8 + 2 = _____

15.　1 + 0 = _____

Count and write the number, and then say it.

1.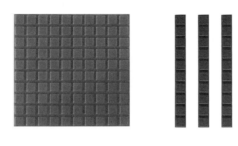

   _____   _____   _____

2.

   _____   _____

Build and say the number.

3. 306   4. 222

Add.

5.   1
    + 1

6.   7
    + 2

7.   9
    + 5

8.   3
    + 8

9.   6
    + 6

10.   7
    + 8

11.   5
    + 4

12.   3
    + 5

13. 5 + 8 = _____

14. 9 + 4 = _____

15. 3 + 3 = _____

16. 6 + 7 = _____

17. 7 + 5 = _____

18. 4 + 8 = _____

Put the numbers in order from the smallest to the largest.

1. 20, 12, 4

20   12   4

———— , ———— , ————

2. 6, 206, 31

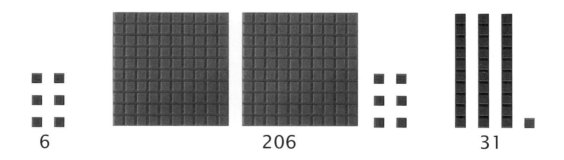

6   206   31

———— , ———— , ————

Put the numbers in order from the largest to the smallest.
Use the blocks.

3. 4, 55, 162

4. 16, 3, 136

_____ , _____ , _____   _____ , _____ , _____

Fill in the blanks with the correct numbers for each sequence.

5. __8__ , _____ , __10__ , __11__ , _____

6. __10__ , __9__ , _____ , _____ , _____

7. __71__ , _____ , _____ , __74__ , _____

Put the numbers in order from the smallest to the largest. Use the blocks.

1. 16, 5, 8

_____ , _____ , _____

2. 25, 100, 63

_____ , _____ , _____

3. 17, 7, 107

_____ , _____ , _____

4. 200, 89, 11

_____ , _____ , _____

Put the numbers in order from the largest to the smallest. Use the blocks.

5. 6, 10, 30

_____ , _____ , _____

6. 20, 80, 10

_____ , _____ , _____

7. 9, 245, 61

_____ , _____ , _____

8. 3, 84, 100

_____ , _____ , _____

Fill in the blanks with the correct numbers for each sequence.

9. __6__ , _____ , _____ , __3__ , _____

10. _____ , __23__ , _____ , __25__ , _____

11. __64__ , _____ , _____ , _____ , __68__

12. _____ , __16__ , __15__ , _____ , _____

LESSON PRACTICE

Put the numbers in order from the smallest to the largest. Use the blocks.

1. 100, 50, 6

_____ , _____ , _____

2. 95, 18, 32

_____ , _____ , _____

3. 10, 400, 40

_____ , _____ , _____

4. 243, 65, 16

_____ , _____ , _____

Put the numbers in order from the largest to the smallest. Use the blocks.

5. 125, 5, 15

_____ , _____ , _____

6. 9, 170, 106

_____ , _____ , _____

7. 22, 99, 48

_____ , _____ , _____

8. 114, 89, 120

_____ , _____ , _____

Fill in the blanks with the correct numbers for each sequence.

9. __29__ , _____ , _____ , __26__ , _____

10. _____ , __44__ , _____ , __46__ , _____

11. __30__ , _____ , _____ , _____ , __34__

12. _____ , __8__ , __7__ , _____ , _____

Put the numbers in order from the smallest to the largest.

1. 17, 15, 5

_____ , _____ , _____

2. 11, 40, 8

_____ , _____ , _____

Put the numbers in order from the largest to the smallest.

3. 75, 12, 43

_____ , _____ , _____

4. 150, 200, 110

_____ , _____ , _____

Fill in the blanks with the correct numbers for the sequence.

5. __79__ , _____ , _____ , __82__ , _____

Build and say the number.

6. 47

7. 109

Add.

8.   9
  + 9

9.   7
  + 8

10.   4
  + 5

11.   2
  + 3

12. 4 + 8 = _____

13. 5 + 7 = _____

14. 3 + 8 = _____

15. Sam had 121 marbles and Jeff had 112 marbles. Who had the larger number of marbles?

_____

16. Riley made four toasted-cheese sandwiches and one peanut butter and jelly sandwich. How many sandwiches did she make in all?

_____

Put the numbers in order from the smallest to the largest.

1. 63, 18, 5

2. 100, 400, 200

_____ , _____ , _____   _____ , _____ , _____

Put the numbers in order from the largest to the smallest.

3. 44, 14, 56

4. 105, 200, 50

_____ , _____ , _____   _____ , _____ , _____

Fill in the blanks with the correct numbers for the sequence.

5. __9__ , _____ , _____ , __6__ , _____

Build and say the number.

6. 98

7. 276

Add.

8.　　6
　　+ 6

9.　　5
　　+ 9

10.　　2
　　+ 7

11.　　4
　　+ 6

12. 8 + 9 = _____

13. 5 + 6 = _____

14. 1 + 7 = _____

15. Marissa picked 7 daisies and 16 roses. What kind of flower did she pick the smaller number of?

_____

16. Jared counted seven trucks and nine cars going by his house this morning. How many vehicles did he count in all?

_____

Put the numbers in order from the smallest to the largest.

1. 39, 19, 99

2. 60, 80, 10

_____ , _____ , _____

_____ , _____ , _____

Put the numbers in order from the largest to the smallest.

3. 299, 18, 74

4. 48, 21,180

_____ , _____ , _____

_____ , _____ , _____

Fill in the blanks with the correct numbers for the sequence.

5. _____ , __89__ , _____ , _____ , __92__

Build and say the number.

6. 240

7. 16

Add.

8.   7
    + 7

9.   2
    + 5

10.   3
    + 7

11.   0
    + 4

12. 8 + 8 = _____

13. 6 + 7 = _____

14. 4 + 9 = _____

15. Emma has 26 school books and 62 story books in her bookcase. What kind of book does she have the larger number of?

_____

16. Kayla lost six pennies and eight dimes. How many coins did she lose in all?

_____

Compare, and then fill in the oval with <, >, or =.

1. 4 $\bigcirc$ 3

2. 5 + 2 $\bigcirc$ 2 + 5

3. 4 + 2 $\bigcirc$ 3 + 1

4. 11 $\bigcirc$ 21

5. 14 $\bigcirc$ 7 + 7

6. 2 + 8 $\bigcirc$ 6 + 3

7. 63 $\bigcirc$ 36

8. 6 + 6 $\bigcirc$ 12

9. 6 + 0 $\bigcirc$ 7

10. 3 + 9 $\bigcirc$ 5 + 5

11. Jennifer's book cost 50 cents and Nelson's book cost 90 cents. Write an inequality showing whose book cost more. (You can write your answer two different ways.)

   _____

12. Denise did five chores for her mom and four chores for her dad. Michael did five chores for his mom and five chores for his dad. Show who did more chores.

   _____

13. Katie read 12 books and Ethan read 8 books. Show who read fewer books.

   _____

14. Chris said that 9 + 5 is the same as 7 + 7. Write a math sentence showing what Chris was saying.

   _____

3B

Compare, and then fill in the oval with <, >, or =.

1. 5 ◯ 7

2. 1 + 1 ◯ 1 + 3

3. 4 + 6 ◯ 2 + 7

4. 13 ◯ 10

5. 18 ◯ 9 + 8

6. 5 + 6 ◯ 8 + 3

7. 27 ◯ 72

8. 7 + 7 ◯ 15

9. 5 + 1 ◯ 6

10. 8 + 4 ◯ 4 + 8

11. William had two rabbits and one duck. His brother had three rabbits and two ducks. Write an inequality showing who had more pets.

   _____

12. Jordan got eight words correct on his spelling test, and Alexa got nine words correct. Write an inequality to show who got more words correct.

   _____

13. Ali said that 6 + 4 is the same as 7 + 3. Write a math sentence, or equation, showing what Ali was saying.

   _____

14. Write an inequality showing which is larger: 105 or 150.

   _____

Compare, and then fill in the oval with <, >, or =.

1. 9 $\bigcirc$ 14

2. 2 + 3 $\bigcirc$ 2 + 4

3. 7 + 2 $\bigcirc$ 5 + 2

4. 25 $\bigcirc$ 15

5. 16 $\bigcirc$ 8 + 8

6. 9 + 6 $\bigcirc$ 9 + 8

7. 31 $\bigcirc$ 13

8. 7 + 5 $\bigcirc$ 12

9. 3 + 4 $\bigcirc$ 5

10. 9 + 2 $\bigcirc$ 3 + 9

11. It is 10 days until Rachel's birthday and 110 days until Jenny's birthday. Write an inequality showing who has fewer days until her birthday.

_____

12. Mom spent 55 dollars on groceries and 65 dollars on clothes for her family. Write an inequality showing where Mom spent more money.

_____

13. Bria said that 9 + 1 is the same as 1 + 9. Write a math sentence, or equation, showing what Bria was saying.

_____

14. Write an inequality showing which is smaller: 213 or 231.

_____

Compare, and then fill in the oval with <, >, or =.

1. 5 + 8 ◯ 3 + 9   2. 57 ◯ 75

Put the numbers in order from the smallest to the largest.

3. 7, 13, 8   4. 90, 50, 70

_____ , _____ , _____   _____ , _____ , _____

Fill in the blanks with the correct numbers for the sequence.

5. __28__ , _____ , _____ , __31__ , _____

Build and say the number.

6. 211   7. 36

Add.

8.   2
    + 9

9.   5
    + 5

10.   3
    + 6

11.   2
    + 8

12. 6 + 9 = ___

13. 3 + 4 = ___

14. 0 + 6 = ___

15. Write an inequality showing which is larger: 452 or 254.

    _____

16. There are five girls and three boys in the Adams family.
     How many children are there altogether?

    _____

Compare, and then fill in the oval with <, >, or =.

1. 2 + 6 $\bigcirc$ 6 + 2   2. 91 $\bigcirc$ 19

Put the numbers in order from the largest to the smallest.

3. 2, 12, 200   4. 105, 15, 51

____ , ____ , ____   ____ , ____ , ____

Fill in the blanks with the correct numbers for the sequence.

5. ____ , __18__ , ____ , ____ , __21__

Build and say the number.

6. 104   7. 13

Add.

8.
$$\begin{array}{r} 4 \\ + 4 \\ \hline \end{array}$$

9.
$$\begin{array}{r} 1 \\ + 6 \\ \hline \end{array}$$

10.
$$\begin{array}{r} 2 \\ + 4 \\ \hline \end{array}$$

11.
$$\begin{array}{r} 0 \\ + 1 \\ \hline \end{array}$$

12. $9 + 9 =$ _____

13. $7 + 8 =$ _____

14. $4 + 7 =$ _____

15. Duncan has three brothers and two sisters. Write an inequality showing whether he has more brothers or more sisters.

_____

16. Alaina walked three miles and rode her bike three miles. How many miles did she travel in all?

_____

Compare, and then fill in the oval with <, >, or =.

1. 1 + 5 ◯ 3 + 4   2. 7 + 8 ◯ 8 + 7

Put the numbers in order from the largest to the smallest.

3. 63, 3, 300   4. 91, 74, 197

_____ , _____ , _____   _____ , _____ , _____

Fill in the blanks with the correct numbers for the sequence.

5. _____ , _____ , _11_ , _____ , _9_

Build and say the number.

6. 135   7. 240

Add.

8.   7
  + 9

9.   2
  + 2

10.   4
  + 9

11.   3
  + 7

12. 1 + 1 = _____

13. 5 + 8 = _____

14. 7 + 7 = _____

15. Sarah said that 9 + 4 is the same as 7 + 6. Write a math sentence, or equation, showing what Sarah was saying.

_____

16. Kiley read six pages in her history book before lunch and eight pages after lunch. How many pages has she read so far?

_____

Answer the questions.

1. Is 44 closer to 40 or 50? _____

2. Is [blocks] closer to [blocks] or [blocks] ? Circle the right answer.

Round to the nearest tens place. The first one is done for you.

3. 25 → 30

4. 58 → _____

Round each number to the nearest ten and estimate the answer. The first one is done for you.

5.   1 8   (2 0)
    + 5 1   (5 0)
                 (7 0)

6.   1 3   (   )
    + 1 2   (   )
                 (   )

7.   2 5   (   )
    + 1 9   (   )
                 (   )

8.   5 1   (   )
    + 4 2   (   )
                 (   )

Add across and down. Then add your answers and see if they match. The first one is done for you.

9.

6	3	9
1	1	2
7	4	11

3	5	
6	1	

10. Erica read 21 books last month and 26 books this month. Estimate how many books she read the last two months.

_____

Answer the questions.

1. Is 38 closer to 30 or 40? _____

2. Is ▌▌ᴇ closer to ▌▌ or ▌▌▌ ? Circle the right answer.

Round to the nearest tens place.

3. 62 → _____   4. 85 → _____

Round to the nearest ten and estimate the answer.

5.　 1 6　(　)
　 + 2 3　(　)
　 ─────　(　)

6.　 4 5　(　)
　 + 3 1　(　)
　 ─────　(　)

7.　 3 4　(　)
　 + 2 2　(　)
　 ─────　(　)

8.　 6 1　(　)
　 + 1 7　(　)
　 ─────　(　)

Add across and down. Then add your answers and see if they match.

9.

3	2	
5	2	

10. When Carol went shopping, she found a coat that cost 58 dollars and a dress that cost 32 dollars. Estimate how many dollars it will cost to buy both.

_____

Answer the questions.

1. Is thirty-one closer to thirty or to forty? _____

2. Is seventy-eight closer to seventy or eighty? _____

Round to the nearest tens place.

3. 72 → _____   4. 55 → _____

Round to the nearest ten and estimate the answer.

5.   2 7   (   )   6.   1 6   (   )
  + 3 2   (___)   + 1 1   (___)
          (   )   (   )

7.   4 4   (   )   8.   2 3   (   )
  + 4 5   (___)   + 3 9   (___)
          (   )   (   )

Add across and down. Then add your answers and see if they match.

9.

4	5	
3	1	

10. It rained 32 inches last year and 26 inches this year. Estimate the total amount of rain that fell in the last two years.

_____

Round to the nearest tens place.

1. 81 → _____

2. 49 → _____

Round to the nearest ten and estimate the answer.

3.   4 6   (   )
    + 2 2   (_____)
              (   )

4.   1 5   (   )
    + 3 3   (_____)
              (   )

Add across and down. Then add your answers and see if they match.

5.

6	1	
2	3	

Compare, and then fill in the oval with <, >, or =.

6. 1 + 3 ◯ 2 + 2

7. 16 ◯ 21

Add.

8.   4
  + 6

9.   1
  + 9

10.   6
  + 9

11.   5
  + 7

12. Kurt worked 15 hours last week and 12 hours this week. Estimate the total time he worked the last two weeks.

_____

13. King John won nine battles. King Henry won eight more battles than King John did. How many battles did King Henry win?

_____

14. Looking out at Grandpa's bird feeder, Jeremiah saw two robins, two bluebirds, and six sparrows. How many birds did he see in all? (First add 2 + 2, and then add 6 to your answer.)

_____

Round to the nearest tens place.

1. 45 → _____

2. 64 → _____

Round to the nearest ten and estimate the answer.

3.
```
    3 6    (      )
  + 1 3    (_____)
           (      )
```

4.
```
    6 7    (      )
  + 1 2    (_____)
           (      )
```

Add across and down. Then add your answers and see if they match.

5.

2	1	
7	0	

Compare, and then fill in the oval with <, >, or =.

6. 5 + 9 ◯ 6 + 7

7. 99 ◯ 105

Add.

8.
$$\begin{array}{r} 3 \\ + 9 \\ \hline \end{array}$$

9.
$$\begin{array}{r} 1 \\ + 8 \\ \hline \end{array}$$

10.
$$\begin{array}{r} 5 \\ + 5 \\ \hline \end{array}$$

11.
$$\begin{array}{r} 2 \\ + 7 \\ \hline \end{array}$$

12. Casey ordered three chocolate and three strawberry ice cream cones for her family. How many cones does she have to carry?

   _____

13. Justin likes to fish. He dug 17 earthworms on Monday and 32 earthworms on Wednesday. Estimate how many worms he has for his next fishing trip.

   _____

14. Last week Isaac's Bagel Shop sold 279 bagels. This week it sold 410 bagels. Write an inequality showing which is the larger number of bagels.

   _____

4F

Round to the nearest tens place.

1. 91 → _____   2. 75 → _____

Round to the nearest ten and estimate the answer.

3.   5 4   (   )   4.   4 9   (   )
     + 3 3   (____)   + 2 1   (____)
              (   )   (   )

Add across and down. Then add your answers and see if they match.

5.

3	3	
2	1	

Compare, and then fill in the oval with <, >, or =.

6. 9 + 9 ◯ 8 + 8   7. 198 ◯ 201

Add.

8.　　9
　　+ 5

9.　　2
　　+ 6

10.　　4
　　+ 3

11.　　3
　　+ 8

12. Haley received nine cards on her birthday. The next day she got two more. How many cards did Haley get in all?

　　_____

13. A salesman sold 53 radios last month and 36 radios this month. Estimate how many he sold in the last two months.

　　_____

14. Write an inequality showing which is smaller: 45 or 405.

　　_____

Build and write using place-value notation.
The first one is done for you.

1.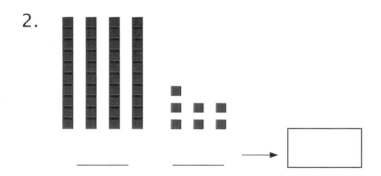

   <u> 200 </u>   <u> 50 </u> <u> 6 </u> ⟶ | 256 |

2.

   _____ _____ ⟶ [   ]

Write using place-value notation.

3. 143 = _____ + _____ + _____

4. 65 = _____ + _____

Add using place-value notation. The first one is done for you.

5. $\begin{array}{r} 2\,4 \\ +\,3\,5 \\ \hline 5\,9 \end{array}$ → $\begin{array}{r} 2\,0+4 \\ 3\,0+5 \\ \hline 5\,0+9 \end{array}$

6. $\begin{array}{r} 6\,2 \\ +\,1\,7 \\ \hline \end{array}$ → $\begin{array}{r} 6\,0+2 \\ 1\,0+7 \\ \hline \end{array}$

7. $\begin{array}{r} 2\,2 \\ +\,2\,3 \\ \hline \end{array}$ → $\begin{array}{r} 2\,0+2 \\ 2\,0+3 \\ \hline \end{array}$

8. $\begin{array}{r} 1\,4\,1 \\ +\,1\,2\,8 \\ \hline \end{array}$ → $\begin{array}{r} 1\,0\,0+4\,0+1 \\ 1\,0\,0+2\,0+8 \\ \hline \end{array}$

9. There are 24 houses on Green Street and 33 houses on Long Lane. How many houses are there altogether?

_____

10. Austin spent 16 dollars at the hardware store and 12 dollars more at the candy store. How much did he spend in all?

_____

Build and write using place-value notation.

1.

_____   _____   _____

2.

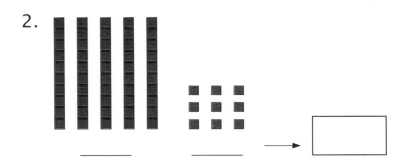

_____   _____

Write using place-value notation.

3. 365 = _____ + _____ + _____

4. 41 = _____ + _____

Add using place-value notation.

5.   64   → 60 + 4   6.   13   → 10 + 3
    + 15 → 10 + 5   + 13 → 10 + 3

7.   57   → 50 + 7
    + 11 → 10 + 1

8.   331   → 300 + 30 + 1
   + 24 →   20 + 4

9. Mom spent 133 dollars at the grocery store and 10 more at the pet store. How much money did Mom spend in all?

_____

10. Twenty-three blackbirds flew over our home yesterday, and twenty-three more flew over today. How many blackbirds flew over in two days?

_____

5C

Build and write using place-value notation.

1.

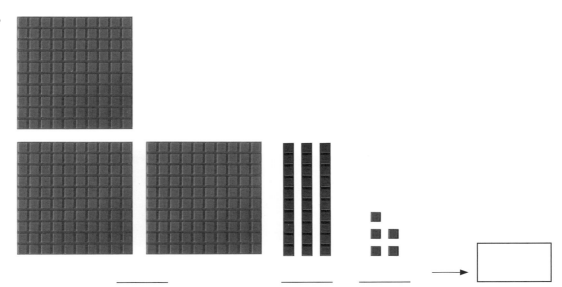

_____   _____   _____   →

2.

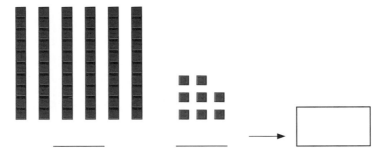

_____   _____   →

Write using place-value notation.

3. 172 = _____ + _____ + _____

4. 27 = _____ + _____

Add using place–value notation.

5.   22   →   20 + 2
      + 22   →   <u>20 + 2</u>

6.   55   →   50 + 5
     + 41   →   <u>40 + 1</u>

7.   89   →   80 + 9
     + 10   →   <u>10 + 0</u>

8.   236   →   200 + 30 + 6
   + 212   →   <u>200 + 10 + 2</u>

9. There are 12 students in our class and 36 students in the other class. How many students would there be if the two classes were combined?

     _____

10. I found 41 cents in my dresser drawer and 46 cents in my desk. How much money did I find altogether?

     _____

Write using place-value notation.

1. 531 = _____ + _____ + _____

2. 18 = _____ + _____

Add using place-value notation.

3.   3 2   →   3 0 + 2
      + 1 1   →   1 0 + 1

4.   1 0 1   →   1 0 0 + 0 0 + 1
      + 3 2 1   →   3 0 0 + 2 0 + 1

Round to the nearest ten and estimate the answer. Then find the exact answer. The first one is done for you.

5.   6 7   (7 0)   6.   4 3   (   )
     + 2 2   (2 0)   + 2 1   (   )
       8 9   (9 0)   (   )

Compare, and then fill in the oval with <, >, or =.

7. 3 + 4 $\bigcirc$ 6 + 6   8. 71 $\bigcirc$ 17

QUICK REVIEW

Solving for the unknown is a good way to review addition and get ready for subtraction.

EXAMPLE $4 + A = 9$ $4 + \overset{5}{A} = 9$

Think, "Four plus what equals nine?"
The answer is five.

Write the answer above the letter that stands for the unknown.

You may use the manipulatives by laying the four bar next to the nine bar and finding what bar is needed to make up the length.

Solve for the unknown.

9. $6 + X = 12$ 10. $7 + R = 11$

11. $5 + B = 7$

12. One hundred twenty geese and two hundred fifty-four swans flew overhead. How many birds did I see?

_____

Write using place-value notation.

1. $114 =$ _____ + _____ + _____

2. $39 =$ _____ + _____

Add using place-value notation.

3.   1 7   →   1 0 + 7
    + 2 2   →   2 0 + 2

4.   2 4 2   →   2 0 0 + 4 0 + 2
    + 1 1 1 →   1 0 0 + 1 0 + 1

Round to the nearest ten and estimate the answer. Then find the exact answer.

5.   6 8   (   )
    + 1 1   (   )
           (   )

6.   5 1   (   )
    + 3 4   (   )
           (   )

Add across and down. Then add your answers and see if they match.

7.

5	6	
7	1	

Compare, and then fill in the oval with <, >, or =.

8. 7 + 8 $\bigcirc$ 8 + 7

9. 103 $\bigcirc$ 331

Solve for the unknown.

10. 7 + X = 10

11. 9 + R = 16

12. 5 + B = 9

13. Tom got 18 questions right on his first test and 21 right on the second. Estimate the total number he got right.

_____

14. We drove 212 miles this morning and 362 miles this afternoon. How many miles did we drive today?

_____

Write using place-value notation.

1. 314 = _____ + _____ + _____

2. 72 = _____ + _____

Add using place-value notation.

3.   1 1   →   1 0 + 1
  + 8 8   →   8 0 + 8

4.   1 7 6   →   1 0 0 + 7 0 + 6
  + 3 2 3   →   3 0 0 + 2 0 + 3

Round to the nearest ten and estimate the answer. Then find the exact answer.

5.   7 2   (   )
  + 1 5   (   )
             (   )

6.   6 1   (   )
  + 1 7   (   )
             (   )

Add across and down. Then add your answers and see if they match.

7.

2	8	
6	3	

Compare, and then fill in the oval with <, >, or =.

8. 2 + 4 $\bigcirc$ 4 + 2

9. 86 $\bigcirc$ 68

Solve for the unknown.

10. 8 + T = 13

11. 4 + Q = 5

12. 9 + Y = 11

13. If Farmer John had 665 cows and 133 calves in the barn, how many animals did he have in the barn?

_____

14. Write an inequality showing which is larger: 21 + 21 or 12 + 12.

_____

Skip count and write the numbers on the lines in the boxes.
Then write the numbers in the spaces underneath.

1.

	2
	___
	6
	___
	___
	12
	___
	___
	___
	20

2.

	___
	4
	___
	8
	10
	___
	___
	___
	___
	___

3.   __2__ , __4__ , _____ , _____ , _____ , _____ , _____ , _____ , _____ , _____

Skip count by two to find each answer.

4. How many cookies? _____

5. How many Xs? _____

6. There are nine children at the party, all wearing their best shoes. How many shoes are there in all?

    _____

Skip count and write the numbers on the lines in the boxes.
Then write the numbers in the spaces underneath.

1.

	2
	___
	___
	8
	___
	___
	___
	16
	18
	___

2.

	___
	4
	6
	___
	10
	___
	___
	___
	___
	___

3. __2__ , _____, _____, __8__ , _____, _____, _____, _____, _____, _____

Skip count by two to find each answer.

4. How many cookies? _____

5. How many Xs? _____

X   X   X   X   X   X   X
X   X   X   X   X   X   X

6. Five elephants are at the zoo. How many great big elephant ears are there?

_____

6C

Skip count and write the numbers on the lines in the boxes.
Then write the numbers in the spaces underneath.

1.

	____
	4
	____
	____
	____
	12
	____
	____
	____
	20

2.

	____
	____
	8
	____
	____
	____
	16
	____
	____

3. __2__ , ____ , ____ , ____ , ____ , ____ , ____ , ____ , __18__ , ____

Skip count by two to find each answer.

4. How many snowflakes? _____

5. Five basketball players lifted both arms to get the rebound. Skip count to find how many arms were in the air.

    _____

6. Ava read two books a week. After three weeks, how many books had Ava read?

    _____

Skip count by two and write the numbers.

1. __2__ , _____ , _____ , __8__ , _____ , _____ , _____ , _____ , _____ , _____

Add using place-value notation.

2.   1 1   →   1 0 + 1
     + 1 1   →   1 0 + 1

3.   1 7 1   →   1 0 0 + 7 0 + 1
     + 1 1 8   →   1 0 0 + 1 0 + 8

Round to the nearest ten and estimate the answer. Then find the exact answer.

4.   1 3   (   )
     + 1 5   (   )
           (   )

5.   3 2   (   )
     + 4 1   (   )
           (   )

Compare, and then fill in the oval with <, >, or =.

6. 5 + 3 ◯ 2 + 4   7. 108 ◯ 801

Solve for the unknown.

8. 5 + X = 8

9. 8 + R = 15

10. 9 + B = 13

11. David remembered to brush his teeth two times a day. Skip count to find how many times he brushed his teeth in one week (seven days).

   _____

12. Sam smiled at his brother seven times, and then he smiled at Debbie two times. Since he was a very happy baby, he smiled three more times. How many times did Sam smile in all?

   _____

Skip count by two and write the numbers.

1. __2__ , _____, _____, _____, _____, _____, _____, _____, _____, __20__

Add using place-value notation.

2.　　3 4　　→　　3 0 + 4
　　+　4　　→　　0 0 + 4

3.　　　3 1 9　　→　　3 0 0 + 1 0 + 9
　　　+ 3 8 0　　→　　3 0 0 + 8 0 + 0

Round to the nearest ten and estimate the answer. Then find the exact answer.

4.　　4 8　　(　　)　　5.　　5 4　　(　　)
　　+ 2 1　　(＿＿＿)　　　　+ 1 2　　(＿＿＿)
　　　　　　(　　)　　　　　　　　(　　)

Compare, and then fill in the oval with <, >, or =.

6. 8 + 2 〇 5 + 3　　　　7. 295 〇 592

Solve for the unknown.

8. $3 + A = 3$

9. $5 + F = 8$

10. $4 + H = 9$

11. Robin spent 126 dollars for groceries the first week and 132 dollars the second week. How much did she spend in all for groceries?

    _____

12. Each of the six deer gave birth to twin fawns. Skip count to find how many fawns were born.

    _____

6F

Skip count by two and write the numbers.

1. _____ , __4__ , _____ , _____ , _____ , _____ , _____ , _____ , __18__ , _____

Add using place-value notation.

2.　　 6 0　　 → 　　 6 0 + 0
　　　+ 　 3　 → 　　 0 0 + 3

3.　　 4 9 2　 → 　　 4 0 0 + 9 0 + 2
　　　+ 2 0 4　 → 　　 2 0 0 + 0 0 + 4

Round to the nearest ten and estimate the answer. Then find the exact answer.

4.　　 1 1　　(　　)
　　　+ 3 3　　(　　)
　　　　　　　(　　)

5.　　 5 3　　(　　)
　　　+ 2 5　　(　　)
　　　　　　　(　　)

Compare, and then fill in the oval with <, >, or =.

6. 6 + 1 〇 1 + 6   7. 21 + 21 〇 12 + 12

Solve for the unknown.

8. $7 + Q = 9$ 9. $6 + X = 12$

10. $7 + Y = 14$

11. Rachel grew two inches every year for four years. Skip count to find how many inches she grew in all.

     _____

12. Evan found 11 pennies in his pocket and 6 pennies under the bed. How many pennies did he find?

     _____

Add using place-value notation and regrouping.
The first one is done for you.

1.
```
  1        1 0
  3 4  →  3 0 + 4
+ 2 8  →  2 0 + 8
  6 2     6 0 + 2
```

2.
```
  4 6  →  4 0 + 6
+ 3 5  →  3 0 + 5
```

3.
```
  7 3  →  7 0 + 3
+ 1 8  →  1 0 + 8
```

4.
```
  3 5  →  3 0 + 5
+ 3 7  →  3 0 + 7
```

5.
```
  2 6  →  2 0 + 6
+ 5 5  →  5 0 + 5
```

6.
```
  3 8  →  3 0 + 8
+ 4 4  →  4 0 + 4
```

Regroup and add.

7.   4 7
    + 2 5

8.   6 6
    + 3 3

9.   7 5
    + 1 6

10. Thomas has two sacks of marbles. One sack has 59 marbles and the other sack has 37 marbles. How many marbles does Thomas have?

_____

Add using place-value notation and regrouping.

1. $\begin{array}{r} 35 \\ +25 \\ \hline \end{array}$ → $\begin{array}{r} 30+5 \\ 20+5 \\ \hline \end{array}$ 2. $\begin{array}{r} 59 \\ +24 \\ \hline \end{array}$ → $\begin{array}{r} 50+9 \\ 20+4 \\ \hline \end{array}$

3. $\begin{array}{r} 45 \\ +28 \\ \hline \end{array}$ → $\begin{array}{r} 40+5 \\ 20+8 \\ \hline \end{array}$ 4. $\begin{array}{r} 76 \\ +15 \\ \hline \end{array}$ → $\begin{array}{r} 70+6 \\ 10+5 \\ \hline \end{array}$

5. $\begin{array}{r} 24 \\ +\ 9 \\ \hline \end{array}$ → $\begin{array}{r} 20+4 \\ +9 \\ \hline \end{array}$ 6. $\begin{array}{r} 25 \\ +39 \\ \hline \end{array}$ → $\begin{array}{r} 20+5 \\ 30+9 \\ \hline \end{array}$

Regroup and add.

7.
```
   2 4
 + 4 8
```

8.
```
   6 7
 +   8
```

9.
```
   5 6
 + 2 4
```

10. Richard and Sarah baked cookies. Sarah baked 26 cookies and Richard baked 38 cookies. How many cookies did they bake altogether?

_____

Add using place-value notation and regrouping.

1. $\begin{array}{r} 2\,5 \\ +\,6\,5 \\ \hline \end{array}$ → $\begin{array}{r} 2\,0+5 \\ 6\,0+5 \\ \hline \end{array}$

2. $\begin{array}{r} 3\,4 \\ +\,4\,5 \\ \hline \end{array}$ → $\begin{array}{r} 3\,0+4 \\ 4\,0+5 \\ \hline \end{array}$

3. $\begin{array}{r} 7\,2 \\ +\,1\,8 \\ \hline \end{array}$ → $\begin{array}{r} 7\,0+2 \\ 1\,0+8 \\ \hline \end{array}$

4. $\begin{array}{r} 6\,8 \\ +\ \ 8 \\ \hline \end{array}$ → $\begin{array}{r} 6\,0+8 \\ +\,8 \\ \hline \end{array}$

5. $\begin{array}{r} 3\,4 \\ +\,4\,6 \\ \hline \end{array}$ → $\begin{array}{r} 3\,0+4 \\ 4\,0+6 \\ \hline \end{array}$

6. $\begin{array}{r} 3\,6 \\ +\,4\,5 \\ \hline \end{array}$ → $\begin{array}{r} 3\,0+6 \\ 4\,0+5 \\ \hline \end{array}$

Regroup and add.

7.
```
   4 8
 +   7
```

8.
```
   2 4
 + 6 6
```

9.
```
   1 8
 + 5 3
```

10. For his birthday, Michael got 16 new baseball cards. He already had 8 cards. How many cards does he have now?

_____

7D

Add. Regroup if needed.

1. $\begin{array}{r} 4\ 9 \\ +\ 4\ 5 \\ \hline \end{array}$

2. $\begin{array}{r} 3\ 6 \\ +\ 3\ 6 \\ \hline \end{array}$

3. $\begin{array}{r} 6\ 8 \\ +\ 2\ 5 \\ \hline \end{array}$

4. $\begin{array}{r} 5\ 5 \\ +\ 1\ 4 \\ \hline \end{array}$

5. $\begin{array}{r} 7\ 7 \\ +\ \ \ 7 \\ \hline \end{array}$

6. $\begin{array}{r} 9\ 5 \\ +\ \ \ 3 \\ \hline \end{array}$

Skip count by two and write the numbers.

7. ____, _4_ , ____, ____, _10_ , ____, ____, ____, ____, ____

Compare, and then fill in the oval with <, >, or =.

8. 9 + 3 $\bigcirc$ 7 + 6

9. 113 $\bigcirc$ 103

Solve for the unknown.

10. 8 + B = 14   11. 6 + G = 10

12. 9 + K = 15

Add across and down. Then add your answers and see if they match.

13.

1	4	
7	3	

14. The clown at Gabriel's party told two jokes a minute. Skip count to find how many jokes the clown told in six minutes.

_____

15. Ian slid down the water slide 14 times, and Jamie slid down 17 times. How many trips down the water slide were made altogether?

_____

7E

Add. Regroup if needed.

1.   1 2
   + 1 3

2.   3 7
   + 2 8

3.   6 3
   + 3 6

4.   2 2
   + 4 8

5.   8 2
   +   9

6.   5 2
   +   6

Skip count by two and write the numbers.

7. _____, _____, __6__, _____, _____, _____, _____, _____, _____, _20_

Round to the nearest tens place.

8. 16 → _____

9. 23 → _____

Solve for the unknown.

10. $9 + B = 14$

11. $5 + G = 6$

12. $7 + K = 13$

Add across and down. Then add your answers and see if they match.

13.

2	6	
5	8	

14. Katherine thinks that 8 + 3 is larger than 6 + 2. Write an inequality showing what Katherine thinks.

_____

15. Mom spent 125 dollars at the grocery store and 172 dollars to have the car fixed. How much money did she spend?

_____

Add. Regroup if needed.

1.
```
  4 3
+ 2 6
```

2.
```
  1 9
+ 1 8
```

3.
```
  6 2
+ 3 2
```

4.
```
  1 5
+ 3 6
```

5.
```
  2 9
+   8
```

6.
```
  7 9
+   6
```

Skip count by two and write the numbers.

7. __2__, _____, _____, _____, _____, _____, _____, _____, _____, _____

Round to the nearest tens place.

8. 48 → _____

9. 65 → _____

Solve for the unknown.

10. 7 + X = 12   11. 8 + W = 11

12. 9 + A = 17

Add across and down. Then add your answers and see if they match.

13.

7	8	
6	5	

14. Krista wrapped eight packages for Christmas. She put two bows on each package. Skip count to find how many bows she used.

    _____

15. Chris counted 28 cows during the car trip, and his sister Kelly counted 13 horses. Estimate how many large animals they saw during the trip, and then find the exact answer.

    _____

8A

Write the correct numbers in the boxes with lines.

1.

										10
										___
										30
										___
										___
										___
										___
										___
										___
										100

Skip count by 10 and write the numbers.

2. __10__ , _____, _____, _____, _____, _____, _____, _____, _____, _____

Skip count by 10 to find how many pennies or cents.

3.   = _____ ¢

4.   = _____ ¢

5. Brad has five dimes. Skip count to find how many cents he has.

   _____

6. Austin would like to buy 30 little toy animals for his farm. If there are 10 animals in a package, how many packages does he need to buy? Skip count to find the answer.

   _____

Write the correct numbers in the boxes with lines.

1.

									_____
									_____
									30
									_____
									_____
									_____
									_____
									_____
									90
									_____

Skip count by 10 and write the numbers.

2. _____, 20, _____, _____, 50, _____, _____, _____, _____, _____

Skip count by 10 to find how many pennies or cents.

3. = _____ ¢

4. = _____ ¢

5. Each ride at the carnival cost 10¢. How much did I spend if I went on eight rides?

    _____

6. Riley has 10 dimes. How many pennies could she change them for?

    _____

Write the correct numbers in the boxes with lines.

1.

									_____
									_____
									_____
									40
									_____
									_____
									_____
									_____
									_____
									100

Skip count by 10 and write the numbers.

2. 10, _____, _____, _____, _____, _____, _____, _____, _____, _____

Skip count by 10 to find how many pennies or cents.

3. = _____ ¢

4. = _____ ¢

5. A candy bar costs 30¢. How many pennies would be needed to buy it?

   _____

6. Joseph made ten dollars a day for six days. Skip count to find how much money he has.

   _____

8D

Skip count by 2 or 10 and write the numbers.

1. _____, 20, _____, 40, _____, _____, _____, _____, _____, _____

2. 2, _____, _____, _____, _____, 12, _____, _____, _____, _____

Add. Regroup if needed.

3.   1 1
     + 3 2

4.   6 4
     + 2 6

5.   5 5
     + 1 7

Add. These do not need regrouping.

6.   3 4 1
     + 1 1 1

7.   6 2 9
     + 2 5 0

8.   1 6 5
     + 5 2 2

Solve for the unknown.

9. $6 + X = 7$

10. $6 + R = 8$

11. $3 + D = 3$

12. Jonathon has two dimes. How many pennies would it take to make the same amount of money?

_____

13. Rachel spent 19 dollars on a new sweater and 36 dollars on a new pair of boots. Estimate how much she spent, and then find the exact amount.

_____

14. James spotted 41 birds yesterday and 48 birds today. How many birds has he seen in the last two days?

_____

15. Which would you rather have, 95 pennies or 9 dimes?

_____

Skip count by 2 or 10 and write the numbers.

1. _____, 4, _____, _____, _____, _____, _____, _____, _____, 20

2. _____, _____, _____, _____, 50, _____, _____, _____, 90, _____

Add. Regroup if needed.

3.   1 5
   + 6 4

4.   1 3
   + 2 8

5.   4 4
   + 4 6

Add. These do not need regrouping.

6.   1 7 5
   +1 1 4

7.   7 3 2
   + 1 5 6

8.   2 4 4
   + 2 3 4

Solve for the unknown.

9. 4 + A = 6

10. 6 + X = 10

11. 9 + F = 18

12. When Jeff was done playing, he put away 6 toys. His room was still a mess, so he put away 2 more. Then his mother told him to put away 3 more toys. How many toys has Jeff put away?

_____

13. Jane had two dimes and her mother gave her five more. How many cents does Jane have?

_____

14. Aunt Lillian has 12 nieces and nephews to make gifts for. She has finished 8 gifts. How many gifts does she still need to make? (Solve for the unknown.)

_____

15. A family of beavers gnawed down 242 trees one day and 157 trees the next day. How many trees did the beavers cut down?

_____

Skip count by 2 or 10 and write the numbers.

1. ____, ____, 30, ____, ____, ____, ____, ____, ____, ____

2. ____, ____, 6, ____, ____, ____, ____, ____, ____, ____

Add. Regroup if needed.

3.　　2 5
　　+ 2 5

4.　　1 9
　　+　 4

5.　　3 8
　　+ 1 5

Add. These do not need regrouping.

6.　　4 3 0
　　+ 2 2 3

7.　　8 0 5
　　+ 1 9 2

8.　　3 1 7
　　+ 6 5 1

Solve for the unknown.

9. $1 + Q = 8$

10. $9 + Y = 17$

11. $4 + T = 5$

12. Joseph cut 56 pieces of red paper and 39 pieces of green paper to make a paper chain for his Christmas tree. How many pieces did he cut?

_____

13. John's mules ate two bales of hay every day. Skip count to find how many bales the mules ate in three days.

_____

14. Zarah has eight pennies and Chance has eight dimes. Which one has more money?

_____

15. Sue worked two hours on Wednesday, two hours on Thursday, and five hours on Friday. How many hours did she work in all?

_____

Write the correct numbers in the boxes with lines.

1.

				10
				15
				50

2.

				5

Skip count by five and write the numbers.

3. 5, _____, _____, _____, _____, _____, _____, _____, _____, _____

Skip count by five to find how many pennies or cents.

4. = _____ ¢

5. Each house has five rooms. Skip count to find how many rooms there are in all.

_____

6. Shelly's mom gave her seven nickels. How many cents does she have?

_____

7. Danica ate five jelly beans after lunch every day. How many jelly beans did she eat in nine days?

_____

Write the correct numbers in the boxes with lines.

1.

				5
				20
				45

2.

				15

Skip count by five and write the numbers.

3. _____, _____, _____, _____, 25, _____, _____, 40, _____, _____

Skip count by five to find how many pennies or cents.

4. = _____ ¢

5. Each shape (pentagon) has five sides. How many sides are there in all?

   _____

6. Oranges cost a nickel each. How much will 9 oranges cost?

   _____

7. Dayna's favorite CD had five songs on it. If she listened to it eight times, how many songs did she hear?

   _____

Write the correct numbers in the boxes with lines.

1.

				25
				40

2.

				10
				35

Skip count by five and write the numbers.

3. 5, _____, _____, _____, _____, _____, _____, _____, _____, 50

Skip count by five to find how many pennies or cents.

4.   = _____ ¢

5. Each flower has five petals. How many petals are there altogether?

    _____

6. Sue has three nickels and Bob has two dimes. Which one has more money?

    _____

7. Dad gave Chris four nickels. How many cents did Dad give Chris?

    _____

Skip count by five or ten and write the numbers.

1. ____, ____, 15, ____, 25, ____, ____, ____, ____, ____

2. ____, ____, ____, ____, 50, ____, 70, ____, ____, ____

Add. Regroup if needed.

3.   6 3
     +   7

4.   2 4
     + 4 8

5.   1 5
     + 4 4

Add. These do not need regrouping.

6.   4 1 2
     + 2 1 6

7.   2 0 3
     + 3 0 2

8.   7 1 3
     + 2 7 2

Solve for the unknown.

9. $6 + X = 8$ 10. $4 + R = 9$

11. $7 + D = 14$

12. Isabella has 40 pennies. Her brother wants to trade 8 nickels for her pennies. Should she say yes?

    _____

13. Five children came in out of the rain. Skip count to find how many feet were tracking mud into the house.

    _____

14. Rodney has 15 comic books and Eugene has 16. How many do they have altogether?

    _____

15. The bus driver picked up six passengers at his first stop and two at his second stop. Then he picked up six more passengers at the third stop. How many passengers has he picked up in all?

    _____

9E

Skip count by five or two and write the numbers.

1. ____, 10, ____, ____, ____, ____, ____, ____, 45, ____

2. ____, ____, ____, 8, ____, ____, 14 , ____, ____, ____

Add. Regroup if needed.

3.
```
   2 7
 + 3 3
```

4.
```
   8 1
 +   3
```

5.
```
   3 6
 + 1 4
```

Add. These do not need regrouping.

6.
```
   2 9 3
 + 1 0 4
```

7.
```
   6 4 5
 + 3 2 1
```

8.
```
   7 8 4
 + 2 1 5
```

Solve for the unknown.

9. $9 + B = 10$

10. $3 + Y = 9$

11. $7 + G = 12$

12. Vontoria needs 45¢ to buy a candy bar. How many nickels does she need?

   _____

13. Playing catch with his dad, Raleigh caught the ball 25 times before lunch and 16 times after lunch. How many times did he catch the ball in all?

   _____

14. Shane's favorite baseball team had 4 runs in the first inning, 2 runs in the second inning, and 8 in the third inning. How many runs did the team have during the first three innings?

   _____

15. Would you rather have five nickels or four dimes?

   _____

Skip count by five or ten and write the numbers.

1. ____, ____, ____, 20, ____, ____, ____, ____, ____, ____

2. ____, 20, ____, ____, ____, ____, ____, ____, ____, ____

Add. Regroup if needed.

3.
```
   5 7
 + 2 2
```

4.
```
   7 4
 +   6
```

5.
```
   2 4
 + 1 8
```

Add. These do not need regrouping.

6.
```
   6 8 0
 + 1 1 9
```

7.
```
   5 3 2
 + 2 2 2
```

8.
```
   1 9 2
 + 2 0 7
```

Solve for the unknown.

9. 8 + R = 10   10. 7 + A = 15

11. 9 + U = 12

Draw lines to match the questions with the right answers.

12. What coin is worth ten cents?

13. What coin is worth one cent?

14. What coin is worth five cents?

15. Mom bought a chair for 314 dollars and a table for 322 dollars. How much did Mom spend on furniture?

   _____

Fill in the blanks and say the amount.

1.

$ _____ . _____ _____

2.

$ _____ . _____ _____

3.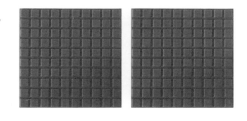

$ _____ . _____ _____

4.

$ _____ . _____ _____

Build and say.

5. $2.62

6. $2.05

7. $1.96

8. $3.18

Fill in the blanks and say the amount.

1.

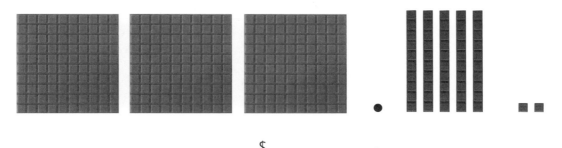

$ _____ . _____ _____

2.

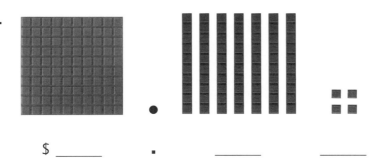

$ _____ . _____ _____

3.

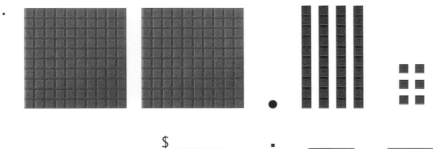

$ _____ . _____ _____

Build and say.

4. $4.51

5. $3.60

6. $1.81

7. $2.07

8. Erica has two dollars, one dime, and five pennies. How much money does she have?

_____

Fill in the blanks and say the amount.

1.

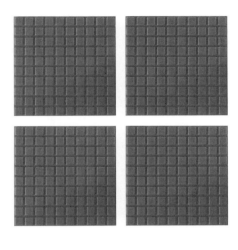

$ _____ . _____ _____

2.

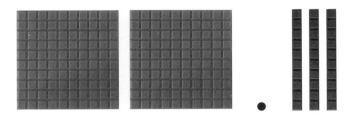

$ _____ . _____ _____

3.

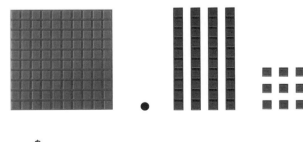

$ _____ . _____ _____

Build and say.

4. $3.23

5. $1.08

6. $4.52

7. $2.90

8. Rhonda found six dollars, seven dimes, and three pennies. How much money did she find?

_____

10D

Fill in the blanks and say the amount.

1.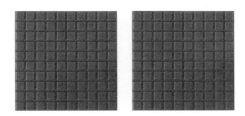

$ _____ . _____ _____

Build and say.

2. $1.48

3. $2.73

4. $4.05

5. $3.60

Skip count to find how many cents.

6. 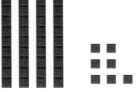 = _____ ¢

Add.

7.　　4 9
　　+　　9

8.　　3 1 1
　　+ 2 3 8

9.　　6 5
　　+ 2 5

Solve for the unknown.

10. 7 + S = 13   11. 3 + B = 5

12. 4 + V = 12

13. Amanda's dad gave her five dollars and four pennies. How much money did she receive?

_____

14. When Jacob went to Disney World, he rode the big roller coaster 19 times and the kiddie roller coaster 25 times. How many roller coaster rides did Jacob have?

_____

15. Elizabeth bought four birthday cards and four get-well cards. The next day she bought nine Christmas cards. How many cards did she buy in all?

_____

10E

Fill in the blanks and say the amount.

1.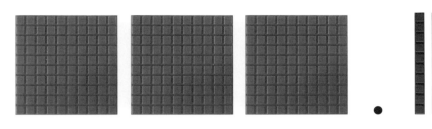

$ _____ . _____ _____

Build and say.

2. $2.31

3. $4.55

4. $1.06

5. $3.78

Skip count to find how many cents.

6.   = ____ ¢

Add.

7.
```
   17
 + 18
```

8.
```
   555
 + 132
```

9.
```
   49
 + 34
```

Round to the nearest tens place.

10. 76 → _____   11. 13 → _____

12. 25 → _____

13. While swinging on the playground, Andrew lost nine nickels from his pocket. Use a decimal point to write how many cents he lost.

   _____

14. On Mother's Day, Aiden bought his mother five white roses and two yellow roses. Then he decided to add two red roses to the bunch. How many roses did he buy in all?

   _____

15. Al has three dimes. How many cents does he have?

   _____

10F

Fill in the blanks and say the amount.

1.  .

$ _____ . _____ _____

Build and say.

2. $1.16

3. $3.09

4. $2.65

5. $4.70

Skip count to find how many cents.

6.   = _____ ¢

Add.

7. 9 2
    + 4
   _____

8. 3 3 7
    +2 0 2
   _____

9. 6 1
    +2 9
   _____

Compare, and then fill in the oval with <, >, or =.

10. 3 + 7 $\bigcirc$ 6 + 1   11. 4 + 4 $\bigcirc$ 8

12. 27 $\bigcirc$ 72

13. Douglas earned eight dollars, six dimes, and nine pennies. How much money did he earn?

_____

14. John swam three miles and ran four miles. The next day he rode his bicycle eight miles. How many miles did he go in all?

_____

15. Lindsay sailed with all her sails up for 29 miles. When the wind picked up, she took in one of the sails and sailed that way for another 18 miles. How many miles did Lindsay sail altogether?

_____

Round to the nearest hundreds place.

1. 190 → _____

2. 206 → _____

3. 355 → _____

Round to the nearest hundred and estimate the answer. Then find the exact answer. The first one is done for you.

```
         1 1
4.      3 6 4      (400)        5.      6 2 8     (        )
      + 2 9 7    + (300)              + 1 7 5    + (        )
      -------    -------              -------      (        )
        6 6 1      (700)
```

```
6.      3 5 9     (        )     7.      5 3 7     (        )
      + 2 5 4    + (        )          + 2 3 3    + (        )
      -------      (        )          -------      (        )
```

```
8.      1 6 8     (        )     9.    1 2 3     (        )
      + 4 5 2    + (        )          + 8 8    + (        )
      -------      (        )          -------      (        )
```

10.   676   (   )   11.   299   (   )
    + 1 4 5   + (   )   + 3 1 1   + (   )
          (   )   (   )

12. First, 124 lights burned out on the big Christmas tree
     downtown. Then 176 more lights burned out. How many
     lights need to be replaced?

     _____

11B

Round to the nearest hundreds place.

1. 476 → _____

2. 515 → _____

3. 610 → _____

Round to the nearest hundred and estimate the answer. Then find the exact answer.

4. $\begin{array}{r} 359 \\ +126 \\ \hline \end{array}$ ( )
  + ( )
    ( )

5. $\begin{array}{r} 138 \\ +212 \\ \hline \end{array}$ ( )
  + ( )
    ( )

6. $\begin{array}{r} 157 \\ +142 \\ \hline \end{array}$ ( )
  + ( )
    ( )

7. $\begin{array}{r} 227 \\ +39 \\ \hline \end{array}$ ( )
  + ( )
    ( )

8. $\begin{array}{r} 449 \\ +137 \\ \hline \end{array}$ ( )
  + ( )
    ( )

9. $\begin{array}{r} 235 \\ +145 \\ \hline \end{array}$ ( )
  + ( )
    ( )

10.　　1 0 9 　　(　　　　)　11.　　4 1 6 　　(　　　　)
　　　+ 2 0 7 　　+ (　　　　)　　　　+ 3 2 9 　　+ (　　　　)
　　　――――　　――――　　　　―――― 　　――――
　　　　　　　　(　　)　　　　　　　　　　(　　)

12. On Monday Steve read 123 pages of his book. On Tuesday
    he read 169 pages. How many pages has he read in all?

    _____

Round to the nearest hundreds place.

1. 450 → _____

2. 103 → _____

3. 278 → _____

Round to the nearest hundred and estimate the answer. Then find the exact answer.

4.   217 (   )
   + 324 + (_____)
          (   )

5.   266 (   )
   + 18 + (_____)
          (   )

6.   134 (   )
   + 365 + (_____)
          (   )

7.   119 (   )
   + 207 + (_____)
          (   )

8.   555 (   )
   + 348 + (_____)
          (   )

9.   806 (   )
   + 106 + (_____)
          (   )

10. 119 (   )   11.   248 (   )
    + 217 + (   )   + 252 + (   )
         (   )   (   )

12. On our trip to Grandma's house, Dad drove 263 miles and Mom drove 179 miles. How far is it to Grandma's house?

    _____

Round to the nearest hundreds place.

1. 755 → _____

2. 115 → _____

Add. Regroup if needed.

3.
```
  8 0 6
+ 1 0 6
```

4.
```
  2 4 8
+ 2 5 2
```

5.
```
  3 3 7
+ 1 7 2
```

6.
```
  5 4
+ 2 8
```

7.
```
  5 3
+ 3 7
```

8.
```
  1 8
+ 2 9
```

Review subtraction facts. These problems review subtracting 0, 1, and 2.

9.
```
  1
- 1
```

10.
```
  1 0
-   2
```

11.   8
    − 1
    ___

12.   3
    − 0
    ___

13.   4
    − 3
    ___

14.   6
    − 2
    ___

15.   5
    − 4
    ___

16.   8
    − 2
    ___

Skip count by two and write the numbers.

17. ____, 4, ____, ____, 10, ____, ____, ____, ____, ____

18. Andrew has five dollars and twenty-six cents. Write that amount with a decimal point and dollar sign.

_____

19. We traveled 55 miles last week and 78 miles this week. How many miles did we travel those two weeks?

_____

20. Jim had $145 in his savings. He got $56 for his birthday. How much money does Jim have now?

_____

Round to the nearest hundreds place.

1. 361 → _____

2. 209 → _____

Add. Regroup if needed.

3.   2 3 5
     + 3 6 5

4.   3 0 0
     + 4 0 9

5.   2 4 9
     + 1 3 2

6.   2 8
     + 3 8

7.   6 5
     + 3 5

8.   5 8
     + 4 2

Review subtraction facts. These problems review subtracting 0, 1, and 2.

9.   4
     − 2

10.   7
     − 2

11.
```
    3
  - 1
```

12.
```
   1 1
  -  2
```

13.
```
    6
  - 5
```

14.
```
    8
  - 0
```

15.
```
   1 0
  -  9
```

16.
```
    9
  - 2
```

Skip count by five and write the numbers.

17. _____, _____, 15, _____, _____, _____, _____, _____, _____, _____

18. A tree grew 17 inches one year and 9 inches the next year. How much did it grow during those two years?

_____

19. Captain Cook spotted 138 penguins in the water and 256 on the shore. How many penguins were spotted by Captain Cook?

_____

20. Mom has five eggs. Two of the eggs are cracked. How many are not cracked? (Watch for word problems that review subtraction.)

_____

Round to the nearest hundreds place.

1. 519 → _____

2. 682 → _____

Add. Regroup if needed.

3.
```
   4 2 9
 + 2 6 6
```

4.
```
   1 0 1
 +   8 9
```

5.
```
   2 3 8
 + 2 4 3
```

6.
```
    9 2
 +    8
```

7.
```
   4 8
 + 3 2
```

8.
```
   6 3
 + 2 7
```

Review subtraction facts. These problems review subtracting by 2 and difference of 2.

9.
```
    5
  - 3
```

10.
```
   1 0
 -   2
```

11.   7
     − 5

12.   6
     − 2

13.   9
     − 7

14.   8
     − 2

15.   1 0
     − 8

16.   1 1
     − 9

Skip count by ten and write the numbers.

17. _____, _____, _____, 40, _____, _____, _____, _____, _____, _____

18. Deb has 21 jelly beans and 48 mints. Round to the nearest 10 and estimate how many pieces of candy she has.

_____

19. Pete drove 512 miles one day and 345 the next day. How far did he drive? Estimate first, and then solve.

_____

20. Drew lost eight pennies. If he finds six, how many are still lost? Write your answer with a decimal point and dollar sign.

_____

Add the money. The first one is done for you.

1.
```
   1 1
  $3. 2 1
+  4. 9 9
  $8. 2 0
```

2.
```
  $7. 0 9
+  1. 9 2
```

3.
```
  $3. 3 3
+  1. 4 4
```

4.
```
  $6. 5 0
+  2. 7 7
```

5.
```
  $4. 0 0
+  2. 5 1
```

6.
```
  $5. 1 9
+  1. 3 8
```

7.
```
  $1. 0 0
+   . 7 5
```

8.
```
  $2. 0 3
+  1. 9 0
```

9.
```
  $8. 7 5
+   . 8 0
```

10. Rose went shopping. If she spent $5.25 in one store and $3.38 in another store, how much did she spend in all?

_____

11. Chance had $2.63 in his pocket. He got $5.50 in a birthday card. How much money does he have now?

_____

12. Joseph bought toys for his pug dog, Max. If he spent $2.99 for a rubber bone and $3.61 for a ball, how much did he spend in all?

_____

12B

Add the money.

1. $7.65
    +   .60

2. $6.31
    + 1.29

3. $5.83
    +   .24

4. $3.19
    +   .90

5. $2.00
    +   .98

6. $1.03
    + 1.25

7. $3.72
    + 4.08

8. $1.99
    + 1.82

9. $2.87
    + 6.89

10. Elizabeth found $3.45 in her drawer and $1.99 behind her dresser. How much money did she find in all?

   _____

11. Meredith wants to buy a book that costs $5.55 and a box of crayons that costs $2.15. How much money does she need?

   _____

12. Daniel's brother gave him $6.34. His sister gave him $2.95. How much money was Daniel given altogether?

   _____

12C

Add the money.

1. $2.13
    + 1.92

2. $4.71
    + 1.36

3. $6.41
    + .39

4. $5.00
    + 2.50

5. $6.63
    + 2.44

6. $7.35
    + 1.05

7. $1.63
    + .72

8. $4.99
    + 3.79

9. $6.33
    + 2.91

10. Caleb lost $5.10 yesterday and $3.91 today. How much did he lose altogether?

   _____

11. Caleb found $2.50 of his lost money. Timothy felt sorry for him and gave him $4.50. How much money does Caleb have now?

   _____

12. Thomas has $3.72 in his left pocket and $3.68 in his right pocket. How much money does he have? Does he have enough to buy a toy that costs $6.00?

   _____

Add. Regroup if needed.

1.
$$\begin{array}{r} \$1.66 \\ +\ 4.08 \\ \hline \end{array}$$

2.
$$\begin{array}{r} \$3.09 \\ +\ 2.56 \\ \hline \end{array}$$

3.
$$\begin{array}{r} \$3.57 \\ +\ 2.62 \\ \hline \end{array}$$

4.
$$\begin{array}{r} 422 \\ +389 \\ \hline \end{array}$$

5.
$$\begin{array}{r} 19 \\ +16 \\ \hline \end{array}$$

6.
$$\begin{array}{r} 17 \\ +25 \\ \hline \end{array}$$

Review subtraction facts. These problems review subtracting 9.

7.
$$\begin{array}{r} 12 \\ -\ 9 \\ \hline \end{array}$$

8.
$$\begin{array}{r} 18 \\ -\ 9 \\ \hline \end{array}$$

9.
$$\begin{array}{r} 9 \\ -\ 9 \\ \hline \end{array}$$

10.
$$\begin{array}{r} 14 \\ -\ 9 \\ \hline \end{array}$$

11. 1 7
     – 9

12. 1 3
     – 9

13. 1 6
     – 9

14. 1 5
     – 9

Fill in the blanks and say the amount.

15.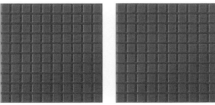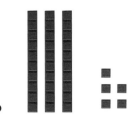

$ _____ . _____ _____

16. Nicholas bought 11 Christmas gifts. If he has wrapped 9 of them, how many are left to wrap?

     _____

17. Forty–six grown-ups and sixty–three children are coming to the wedding. Estimate how many chairs are needed, and then add to find the exact number.

     _____

18. Jacob spent $2.78 for a sandwich and $1.19 for a drink. How much did he spend in all?

     _____

Add. Regroup if needed.

1. $1. 6 8
    + 4. 7 7

2. $4. 5 6
    + 4. 4 4

3. $2. 6 3
    + . 5 1

4. 6 8 4
    + 1 2 2

5. 6 2
    + 2 9

6. 8 3
    + 7

Review your subtraction facts. These problems review subtracting 8.

7. 1 1
    − 8

8. 9
    − 8

9. 1 7
    − 8

10. 1 2
    − 8

11.
   $\begin{array}{r} 1\ 4 \\ -\ 8 \\ \hline \end{array}$

12.
   $\begin{array}{r} 1\ 3 \\ -\ 8 \\ \hline \end{array}$

13.
   $\begin{array}{r} 1\ 5 \\ -\ 8 \\ \hline \end{array}$

14.
   $\begin{array}{r} 1\ 6 \\ -\ 8 \\ \hline \end{array}$

Fill in the blanks and say the amount.

15.

$ _____ . _____ _____

16. Emily had 13 dimes, but she lost 8 of them. How many dimes does she have left? How many cents is that?

17. Ava wrapped 267 candies with nuts and 197 plain candies to sell in her dad's shop. How many candies did she wrap in all? Round to hundreds and estimate, and then solve.

_____

18. Michael has 3 dimes and 5 nickels. Find out how much money he has in all and write the answer with a dollar sign and decimal point.

_____

12F

Add. Regroup if needed.

1.
$$\begin{array}{r} \$2.78 \\ +\ 6.58 \\ \hline \end{array}$$

2.
$$\begin{array}{r} \$3.52 \\ +\ 1.77 \\ \hline \end{array}$$

3.
$$\begin{array}{r} \$8.91 \\ +\ .05 \\ \hline \end{array}$$

4.
$$\begin{array}{r} 379 \\ +264 \\ \hline \end{array}$$

5.
$$\begin{array}{r} 54 \\ +18 \\ \hline \end{array}$$

6.
$$\begin{array}{r} 47 \\ +\ 9 \\ \hline \end{array}$$

Review subtraction facts. These problems review subtracting 8 and the doubles.

7.
$$\begin{array}{r} 12 \\ -\ 6 \\ \hline \end{array}$$

8.
$$\begin{array}{r} 13 \\ -\ 8 \\ \hline \end{array}$$

9.
$$\begin{array}{r} 10 \\ -\ 5 \\ \hline \end{array}$$

10.
$$\begin{array}{r} 14 \\ -\ 7 \\ \hline \end{array}$$

11.
```
  1 2
-   8
```

12.
```
  1 7
-   8
```

13.
```
   8
-  4
```

14.
```
   6
-  3
```

Fill in the blanks and say the amount.

15.

$. _____ _____

16. Sixteen birds landed on my feeder. Eight birds flew away. How many birds are left on the feeder?

_____

17. Bria has three nickels and five pennies. Use a dollar sign and decimal point to write how much money she has in all.

_____

18. Kara talked on the phone 35 minutes yesterday and 17 minutes today. How many minutes did she talk in all?

_____

Add the columns. Find 10 first if you can.

1.　　2
　　　　8
　　＋ 5

2.　　6
　　　　3
　　＋ 4

3.　　5
　　　　4
　　＋ 5

4.　　8
　　　　2
　　　　6
　　＋ 1

5.　　6
　　　　2
　　　　4
　　＋ 7

6.　　3
　　　　7
　　　　5
　　＋ 5

Add the columns. Find 10 first if you can.

7.
```
   4 0
   5 0
 + 1 0
```

8.
```
   3 0
   7 0
 + 4 0
```

9.
```
   5 1
   5 9
 + 2 0
```

Add. First look for 10.

10. $9 + 1 + 8 + 2 =$ _____

11. $6 + 1 + 8 + 4 + 2 =$ _____

12.   Bill got 4 presents for his birthday from his grandparents, 5 from his parents, 6 from his brothers and sisters, and 5 from his friends. How many presents did he get in all?

_____

Add the columns. Find 10 first if you can.

1.
```
    6
    4
  + 9
_____
```

2.
```
    9
    5
  + 5
_____
```

3.
```
    7
    2
  + 3
_____
```

4.
```
    8
    5
    4
  + 2
_____
```

5.
```
    9
    1
    7
  + 3
_____
```

6.
```
    1
    2
    3
  + 9
_____
```

7.
```
    6 0
    4 0
  + 2 0
```

8.
```
    2 0
    6 0
  + 8 0
```

9.
```
    4 3
    2 7
  + 6 1
```

Add. First look for 10.

10.   8 + 2 + 3 + 7 + 2 = _____

11.   4 + 9 + 6 + 2 + 1 = _____

12.   For gym class we ran laps around the field. Jason ran 8 laps, Mary ran 2 laps, Susan ran 3 laps, and Eric ran 3 laps. How many laps were run in all?

_____

Add the columns. Find 10 first if you can.

1.   2
      8
  + 7

2.   5
      1
  + 5

3.   4
      3
  + 3

4.   2
      8
      2
  + 3

5.   3
      8
      7
  + 2

6.   9
      5
      6
  + 1

7.
```
    8 0
    5 0
    1 0
  + 1 0
  ------
```

8.
```
    4 0
    6 0
      4
  +   2
  ------
```

9.
```
    8 4
    2 6
    1 7
  + 2 3
  ------
```

Add. First look for 10.

10. $6 + 7 + 3 + 4 + 6 =$ _____

11. $5 + 5 + 4 + 3 + 7 =$ _____

12. Mrs. Green's class had a reading contest. Here are the numbers of books read by the different students: 11, 14, 5, 10, and 4. How many books were read altogether?

_____

Add the columns. Find 10 first if you can.

1.   3
       7
  + 6

2.   3
       9
       1
  + 7

3.   4 2
     6 4
     8 2
   + 2

4. $2. 8 5
  + 6. 5 6

5.   1 4 9
  + 2 7 3

6.   1 4
   + 9

Review subtraction facts. These problems review making 9 and 10.

7.   1 0
   − 8

8.   1 0
   − 4

9.   9
   − 6

10.   1 0
   − 7

11.   9
   − 2

12.   9
   − 4

13.   9
      – 1

14.   1 0
      – 6

Skip count and write the numbers.

15. _____, 20, _____, 40, _____, _____, _____, _____, _____, _____

QUICK REVIEW

Remember the names of these three different shapes.

 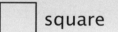triangle   □ square   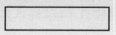rectangle

16. Circle the name of the shape with three sides.

    square   rectangle   triangle

17. The doctor helped 13 patients on Wednesday, 15 on
    Thursday, and 11 on Friday. How many patients did he help
    those three days?

    _____

18. Ruth has $6.18 and her mother gave her $2.00. Does she
    have enough money to buy a toy that costs $9.00?

    _____

Add the columns. Find 10 first if you can.

1.
```
   2
   4
 + 2
─────
```

2.
```
   6
   3
   8
 + 2
─────
```

3.
```
  1 0
  1 0
  2 0
+ 2 0
─────
```

4.
```
  $3. 4 6
+   2. 5 4
──────────
```

5.
```
  1 0 0
+ 2 7 8
────────
```

6.
```
  7 9
+ 8 8
──────
```

Review subtraction facts. These problems review making 9 and 10.

7.
```
  1 0
 -  2
─────
```

8.
```
  1 0
 -  5
─────
```

9.
```
   9
 - 5
─────
```

10.
```
   9
 - 7
─────
```

11.
```
  1 0
 -  3
─────
```

12.
```
   9
 - 8
─────
```

13.
```
   9
 - 3
 ___
```

14.
```
  1 0
 - 4
 ___
```

Skip count and write the numbers.

15. ____, ____, 6, ____, 10, ____, ____, ____, ____, ____

16. Circle the name of the shape with four sides.

square   triangle   rectangle

17. Julia had an ice cream shop. She sold 35 small cones, 22 large cones, and 12 extra large cones. How many ice cream cones did she sell in all?

_____

18. David has nine nickels and seven pennies. How much money does he have?

_____

Add the columns. Find 10 first if you can.

1.　　5
　　　　5
　　　+ 2

2.　　5
　　　　6
　　　　4
　　　+ 7

3.　　2 3
　　　　3 7
　　　　2 3
　　　+ 1 2

4.　　$4. 3 6
　　　+　1. 2 2

5.　　1 1 6
　　　+ 6 8

6.　　1 6
　　　+ 4 4

Review subtraction facts. These may be any of the facts reviewed so far.

7.　1 7
　　－ 8

8.　1 4
　　－ 7

9.　1 0
　　－ 3

10.　9
　　－ 5

11.　1 5
　　－ 9

12.　9
　　－ 7

13.
```
  1 2
-   8
```

14.
```
  1 6
-   9
```

Skip count and write the numbers.

15. _____, _____, 15, _____, _____, _____, _____, _____, _____, 50

Match the shapes with their names.

16.   rectangle

17.   triangle

18.   square

19. Dan has six nickels, five dimes, and four pennies. How much money does he have altogether?

_____

20. Timothy counted 33 raisins in his cereal bowl, and Peter counted 49 in his bowl. How many raisins do the boys have in all?

_____

Measure using the two block, which is one inch long.

1.

_____ inches   _____ inches

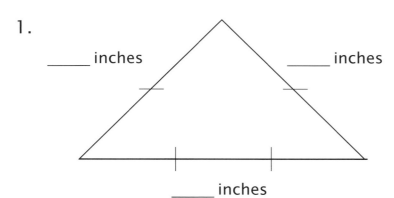

_____ inches

2.

_____ inches

Measure using a ruler.

3.

_____ "

4.

_____ "

5.

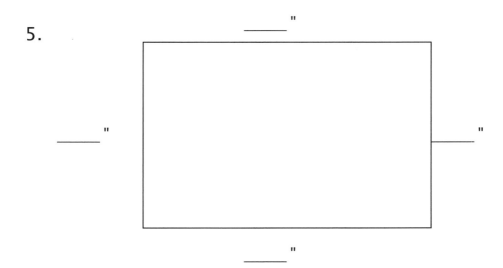

6. 12 inches = _____ foot.

14B

Measure using the two block, which is one inch long.

1. _____ inch

_____ inch   _____ inch

_____ inch

2. _____ inches

Measure using a ruler.

3.

_____ "

4.

_____ "

5.

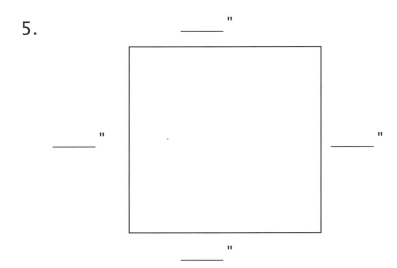

6. _____ inches = 1 foot.

Measure using the two block, which is one inch long.

1.

_____ inches

_____ inch

_____ inch

_____ inches

2.

_____ inches

Measure using a ruler.

3.

_____ "

4.

_____ "

5.

_____"

_____"   _____"

_____"

6. 2 feet = _____ inches. (Add 12 + 12 to find the answer.)

Measure using a ruler.

1.

_____"

2.

_____"

Add.

3.
$$\begin{array}{r} \$1.77 \\ +\ 2.78 \\ \hline \end{array}$$

4.
$$\begin{array}{r} 118 \\ +122 \\ \hline \end{array}$$

5.
$$\begin{array}{r} 23 \\ +24 \\ \hline \end{array}$$

6. $1 + 2 + 3 + 7 + 8 + 9 =$ _____

7. $6 + 6 + 4 + 3 + 7 =$ _____

Compare, and then fill in the oval with <, >, or =.

8. 10 – 4 $\bigcirc$ 3 + 3   9. 8 – 2 $\bigcirc$ 5 + 3

Subtraction review. Subtraction may be written horizontally.

10. 7 – 3 = _____   11. 8 – 5 = _____   12. 7 – 4 = _____

13. 8 – 3 = _____   14. 9 – 5 = _____   15. 15 – 9 = _____

16. How many sides does a triangle have?

_____

17. Jared has six dimes. If he lost three dimes, how many cents does he have left?

_____

18. Mom bought 6 apples, 5 oranges, 10 bananas, and 4 pears. How many pieces of fruit did she buy?

_____

Measure using a ruler.

1. ____"

2. ____"

Add.

3.
$$\begin{array}{r} \$3.\,6\,3 \\ +\ \ 2.\,7\,7 \\ \hline \end{array}$$

4.
$$\begin{array}{r} 4\,5\,2 \\ +\,1\,8\,1 \\ \hline \end{array}$$

5.
$$\begin{array}{r} 8\,9 \\ +\,8\,7 \\ \hline \end{array}$$

6. $3 + 3 + 7 + 3 + 2 =$ ____

7. $10 + 2 + 10 + 2 =$ ____

Compare, and then fill in the oval with <, >, or =.

8. 7 – 3 $\bigcirc$ 7 – 4

Subtraction review.

9. 11 – 7 = _____   10. 13 – 7 = _____   11. 8 – 3 = _____

12. 16 – 7 = _____   13. 12 – 7 = _____   14. 15 – 7 = _____

15. How many sides does a rectangle have?

_____

16. Bonnie flew 256 miles in her airplane. After she landed to get fuel, she took off again and flew another 289 miles. How far did she fly in all?

_____

17. Peter called Debbie three times, his mother four times, the doctor one time, and the library two times. How many calls did Peter make in all?

_____

18. Sam is three feet tall. How many inches tall is he? (Use column addition.)

_____

Measure using a ruler.

1.
_____ "

2.
_____ "

Add.

3.
```
  $1.5 6
+   2.3 8
```

4.
```
    4 1 9
 +  4 1 9
```

5.
```
   3 9
 + 4 2
```

6. 5 + 5 + 5 + 5 = _____

7. 3 + 5 + 10 = _____

Compare, and then fill in the oval with <, >, or =.

8. 14 – 9 ◯ 4 + 4

Subtraction review.

9. 11 – 6 = _____   10. 14 – 6 = _____   11. 15 – 6 = _____

12. 13 – 6 = _____   13. 15 – 7 = _____   14. 7 – 4 = _____

15. How many sides does a square have?

_____

16. Matthew has seven dimes and his sister has three nickels. How much money do they have together?

_____

17. Bill's book has 13 pages. He read three pages before lunch and three pages after lunch. How many pages does he have left to read?

_____

18. Jerry traveled 56 miles the first day and 35 miles the second day. How many miles did Jerry travel in those two days?

_____

15A

Circle the name of each shape. Count the spaces on each side and write the distance in the box. Add up the peRIMeter (distance around) each shape. Notice that we have drawn the inches smaller than they really are. The first one is done for you.

1. The shape is a

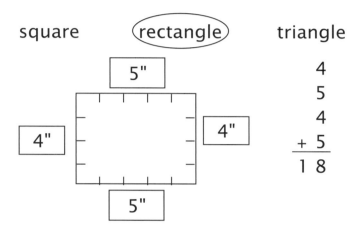

square   (rectangle)   triangle

```
  4
  5
  4
+ 5
-----
1 8
```

The perimeter is ___18___ inches.

2. The shape is a

square   rectangle   triangle

The perimeter is _____ inches.

3. The shape is a:

    square   rectangle   triangle

    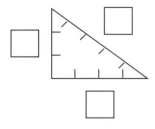

    The perimeter is _____ inches.

4. The shape is a:

    square   rectangle   triangle

    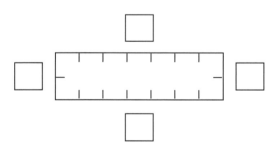

    The perimeter is _____ inches.

Circle the name of each shape. Count the spaces on each side and write the distance in the box. Add up the peRIMeter (distance around) each shape.

1. The shape is a

    square   rectangle   triangle

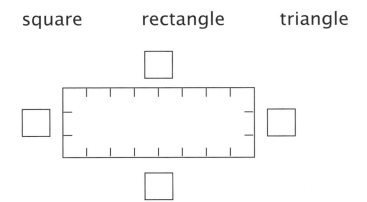

    The perimeter is _____ inches.

2. The shape is a

    square   rectangle   triangle

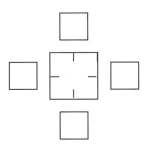

    The perimeter is _____ inches.

3. The shape is a

square   rectangle   triangle

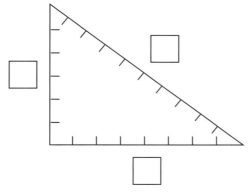

The perimeter is _____ inches.

4. The shape is a

square   rectangle   triangle

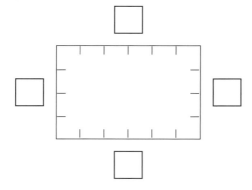

The perimeter is _____ inches.

Circle the name of each shape. Add to find the perimeter.
The distance is given for you.

1. The shape is a

   square   rectangle   triangle

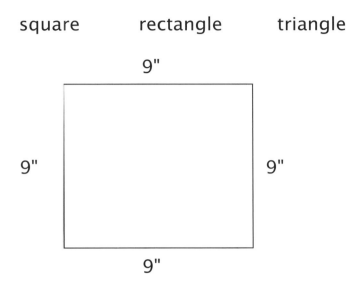

9"

9"   9"

9"

   The perimeter is _____ inches.

2. The shape is a

   square   rectangle   triangle

   5"

2" [   ] 2"

   5"

   The perimeter is _____ inches.

3. The shape is a

   square   rectangle   triangle

   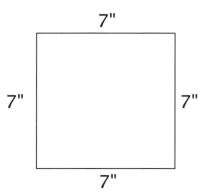

   The perimeter is _____ inches.

4. The shape is a

   square   rectangle   triangle

   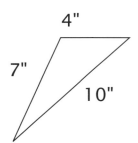

   The perimeter is _____ inches.

Circle the name of each shape. Add to find the perimeter. The distance is given for you.

1. The shape is a

   square   rectangle   triangle

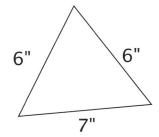

   The perimeter is _____ inches.

2. The shape is a

   square   rectangle   triangle

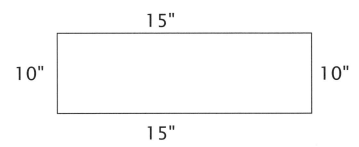

   The perimeter is _____ inches.

Add.

3. $2. 4 9
     + 1. 3 2

4.   3 0 0
     + 4 0 9

5.   2 7
     + 2 5

Subtraction review.

6. 11 – 5 = _____   7. 12 – 4 = _____   8. 13 – 5 = _____

9. 11 – 3 = _____   10. 12 – 5 = _____   11. 11 – 4 = _____

12. Ashley is seven years old. How many years will go by before she is ten years old?

    _____

13. Anthony sold 53 newspapers in the morning and 8 more in the evening. How many papers did he sell in all?

    _____

14. Luke drew a line two feet long. How many inches long was the line?

    _____

15. Seth has nine nickels and one dime. How many cents does he have?

    _____

Circle the name of the shape. Add to find the perimeter.

1. The shape is a

   square   rectangle   triangle

   6"

   6"   6"

   6"

   The perimeter is _____ inches.

2. The shape is a

   square   rectangle   triangle

   The perimeter is _____ inches.

Add.

3. $4. 2 8   4.   2 8 5   5.   4 5
   + 1. 6 5   + 1 5 6   + 5 5

Add across and down. Then add your answers and see if they match.

6.

7	9	
3	5	

7.

6	4	
9	8	

Subtraction review.

8. 14 − 5 = _____

9. 12 − 3 = _____

10. 6 − 1 = _____

11. 13 − 4 = _____

12. Ron drew a square that was four inches long on each side. What was the perimeter of the square?

_____

13. January and March each have 31 days. February has only 28 days. How many days are there in January, February, and March?

_____

14. Katie is four feet tall. How many inches tall is Katie?

_____

Circle the name of the shape. Add to find the perimeter.

1. The shape is a

    square   rectangle   triangle

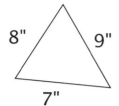

    The perimeter is _____ inches.

2. The shape is a

    square   rectangle   triangle

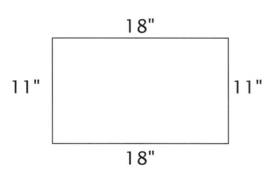

    The perimeter is _____ inches.

Add.

3. $2.48   4.   1 8 3   5.   6 3
    + 1.29   +   7 7   + 2 6

Skip count and write the numbers.

6. 2, ____, ____, ____, ____, ____, ____, ____, ____, ____

7. 5, ____, ____, ____, ____, ____, ____, ____, ____, ____

Subtraction review.

8. 9 – 6 = ____

9. 11 – 7 = ____

10. 12 – 5 = ____

11. 13 – 9 = ____

12. Ava invited 11 people to her birthday party. Five of them have already arrived. How many guests is she waiting for?

_____

13. Donna made a garden shaped like a triangle. The sides of the triangle were six feet, eight feet, and ten feet. How many feet of fence does she need to go all around her garden? (This is a perimeter problem.)

_____

14. The Pirates scored nine runs in the first inning and nine runs in the second inning. What is their score so far?

_____

Write the number and say it. The first one is done for you.

1. $1,000 + 400 + 10 + 8 = \underline{1,418}$

2. $30,000 + 5,000 + 300 + 60 + 1 = \underline{\hspace{2cm}}$

3. $700,000 + 80,000 + 5,000 + 800 + 90 + 2 = \underline{\hspace{2cm}}$

4. $200,000 + 60,000 + 5,000 + 100 + 40 + 3 = \underline{\hspace{2cm}}$

5. $6,000 + 200 + 30 + 7 = \underline{\hspace{2cm}}$

Write using place-value notation. The first one is done for you.

6. $13,129 = \underline{10,000} + \underline{3,000} + \underline{100} + \underline{20} + \underline{9}$

7. $2,356 = \underline{\hspace{1.5cm}} + \underline{\hspace{1.5cm}} + \underline{\hspace{1.5cm}} + \underline{\hspace{1.5cm}}$

Write the number and say it. The first one is done for you.

8. three hundred forty-five thousand, four hundred eighty-two =

   <u>345,482</u>

9. one thousand, five hundred forty-two = _____

Add. The answers will include the thousands place. The first one is done for you.

10.
```
      1
     8 6 3
  +    3 5 1
  _____
   1, 2 1 4
```

11.
```
     9 1 5
  + 4 3 6
  _____
```

12.
```
    3 8 1
  + 7 2 7
  _____
```

Write the number and say it.

1. $2{,}000 + 700 + 90 + 4 =$ _____

2. $10{,}000 + 6{,}000 + 300 + 20 + 2 =$ _____

3. $600{,}000 + 50{,}000 + 1{,}000 + 700 + 40 + 1 =$ _____

4. $500{,}000 + 30{,}000 + 6{,}000 + 500 + 80 + 3 =$ _____

5. $2{,}000 + 500 + 40 + 9 =$ _____

Write using place-value notation.

6. $41{,}456 =$ _____ + _____ + _____ + _____ + _____

7. $238{,}199 =$ _____ + _____ + _____ + _____ + _____ + _____

Write the number and say it.

8. three thousand, one hundred twenty-one = _____

9. forty-five thousand, six hundred sixteen = _____

Add. The answers will include the thousands place.

10.
```
    5 9 3
  + 5 5 1
```

11.
```
    8 7 6
  + 4 3 1
```

12.
```
    9 6 7
  + 2 0 2
```

Write the number and say it.

1. 1,000 + 200 + 20 + 4 = _____

2. 40,000 + 3,000 + 600 + 30 + 8 = _____

3. 200,000 + 40,000 + 7,000 + 500 = _____

4. 100,000 + 20,000 + 2,000 + 400 + 70 + 2 = _____

5. 7,000 + 200 + 90 + 4 = _____

Write using place-value notation.

6. 56,644 = _____ + _____ + _____ + _____ + _____

7. 3,256 = _____ + _____ + _____ + _____

Write the number and say it.

8. one thousand, eight hundred thirty-eight = _____

9. thirty-three thousand, two hundred thirty = _____

Add. The answers will include the thousands place.

10.
```
   4 5 3
 + 7 1 4
```

11.
```
   3 4 5
 + 9 7 8
```

12.
```
   7 1 6
 + 5 6 3
```

Write the number and say it.

1. $4,000 + 800 + 10 + 9 =$ _____

2. $50,000 + 7,000 + 200 + 80 + 4 =$ _____

Circle the name of the shape. Add to find the perimeter.

3. The shape is a

square   rectangle   triangle

32"

11"   11"

32"

The perimeter is _____ inches.

Add.

4. 9 0 1
   + 8 5 0

5. 5 3 4
   + 6 7 3

6.
$$\begin{array}{r} \$4.\,1\,2 \\ +\quad 4.\,7\,1 \\ \hline \end{array}$$

7.
$$\begin{array}{r} 1\,7 \\ 2\,3 \\ +\,5\,5 \\ \hline \end{array}$$

Subtraction review.

8. $10 - 6 =$ _____

9. $15 - 8 =$ _____

10. $14 - 6 =$ _____

11. $6 - 3 =$ _____

12. $12 - 7 =$ _____

13. $6 - 4 =$ _____

14. Abigail went shopping. If she spent $48 the first day, $32 the second day, and $21 the third day, how much did she spend in all?

_____

15. Lisa drew a line three feet long. How many inches long was the line?

_____

16. John is 16 and Justin is 8. What is the difference in their ages?

_____

Write the number and say it.

1. 6,000 + 200 + 10 + 1 = _____

2. twenty-eight thousand, six hundred sixteen = _____

Circle the name of the shape. Add to find the perimeter.

3.   The shape is a

square   rectangle   triangle

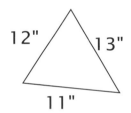

The perimeter is _____ inches.

Add.

4.   5 5 9
     + 5 2 4
     _____

5.   9 4 3
     + 4 7 5
     _____

6.
$$\begin{array}{r} \$3.\,37 \\ +\quad 8.\,29 \\ \hline \end{array}$$

7.
$$\begin{array}{r} 6\,5 \\ 2\,8 \\ +\;4\,5 \\ \hline \end{array}$$

Subtraction review.

8. $9 - 3 =$ _____

9. $8 - 5 =$ _____

10. $14 - 8 =$ _____

11. $11 - 4 =$ _____

12. $16 - 7 =$ _____

13. $7 - 1 =$ _____

14. Molly bought two pieces of furniture for her new house. They cost $198 and $242. Round to the nearest hundred and estimate. Then find exactly what she spent.

_____

15. Jim had eight dimes. He gave his little brother 30¢. How much money does Jim have left?

_____

16. A garden is shaped like a rectangle. It is nine feet long and seven feet wide. How many feet long does the fence need to be (perimeter)?

_____

Write the number and say it.

1. 7,000 + 800 + 50 + 4 = _____

2. eight hundred fifteen thousand, two hundred thirty-one =

    _____

Circle the name of the shape. Add to find the perimeter.

3.   The shape is a

square   rectangle   triangle

6"

6" ☐ 6"

6"

The perimeter is _____ inches.

Add.

4.   6 2 4
    + 4 1 3

5.   4 2 6
    + 8 7 3

6.
```
  $1.32
+  3.38
```

7.
```
   71
   53
+  11
```

Subtraction review.

8. 8 – 7 = _____

9. 11 – 8 = _____

10. 14 – 9 = _____

11. 9 – 4 = _____

12. 13 – 5 = _____

13. 8 – 3 = _____

14. James met 512 cars on his way to work and 471 cars on his way home. How many cars did he meet that day? Estimate first, and then solve.

_____

15. Christina needs 10 apples to make a pie. If she has 7 apples now, how many more does she need?

_____

16. Would you rather have five nickels or four dimes?

_____

17A

Round to the nearest thousands place.

1. 4,850 → _____   2. 1,324 → _____

Round to the nearest ten thousands place.

3. 87,416 → _____   4. 52,933 → _____

Round to the nearest thousand and estimate the answer. Then find the exact answer. The first one is done for you.

```
      1  1
5.    7, 3 7 3          ( 7,000  )
    + 4, 6 6 1        + ( 5,000  )
    1 2, 0 3 4          (12,000   )
```

```
6.    4, 8 5 9          (          )
    + 2, 4 4 4        + (          )
                        (          )
```

```
7.    9, 2 5 3          (          )
    + 7, 8 4 5        + (          )
                        (          )
```

8.
```
    7, 1 3 2
  + 1, 1 8 6
```

(   )
+ (_____)
(   )

9.
```
    3, 6 2 4
  + 4, 4 1 8
```

(   )
+ (_____)
(   )

10.
```
    2, 8 5 2
  + 3, 1 4 9
```

(   )
+ (_____)
(   )

11. Sue traveled 3,152 miles this week and 7,321 miles last week. How far did she travel in all?

_____

12. 5,232 big fish and 3,765 little fish swam up the stream. How many fish were there altogether?

_____

17B

Round to the nearest thousands place.

1. 7,056 → _____

2. 3,512 → _____

Round to the nearest ten thousands place.

3. 45,346 → _____

4. 21,918 → _____

Round to the nearest thousand and estimate the answer. Then find the exact answer.

5.   5, 2 4 2   (   )
    + 3, 7 6 5   + (   )
                        (   )

6.   9, 2 8 7   (   )
    + 1, 3 2 1   + (   )
                        (   )

7.   6, 4 6 3   (   )
    + 8, 7 6 5   + (   )
                        (   )

8. 7, 2 1 4
    + 1, 1 0 8
           (   )
    + (_____)
           (   )

9. 2, 8 1 7
    + 9, 2 3 6
           (   )
    + (_____)
           (   )

10. 3, 6 8 0
    + 3, 3 8 4
           (   )
    + (_____)
           (   )

11. Christina planted 5,740 carrot seeds and 4,291 celery seeds. How many seeds did she plant?

_____

12. Debbie counted 4,987 stars in one part of the sky and 3,732 stars in another part of the sky. How many did she count altogether?

_____

17C

Round to the nearest thousands place.

1. 1,476 → _____   2. 5,746 → _____

Round to the nearest ten thousands place.

3. 79,813 → _____   4. 14,375 → _____

Round to the nearest thousand and estimate the answer. Then find the exact answer.

5.   9, 4 1 3   (   )
      + 1, 2 4 5   + (   )
                          (   )

6.   9, 2 8 7   (   )
      + 7, 4 9 1   + (   )
                          (   )

7.   5, 4 8 6   (   )
      + 4, 5 2 8   + (   )
                          (   )

8.
```
   9, 0 2 5
 + 3, 3 5 4
```
```
  (            )
+ (            )
  (            )
```

9.
```
   7, 5 1 3
 + 7, 2 5 4
```
```
  (            )
+ (            )
  (            )
```

10.
```
   1, 8 9 0
 + 3, 6 7 2
```
```
  (            )
+ (            )
  (            )
```

11. Susan bought a used car for $5,486. Then she spent $1,194 on repairs. How much did she spend altogether on her car?

   _____

12. There are 8,972 people living in Greenville. Last night 9,221 more people came to watch the parade. How many people were in Greenville last night?

   _____

Round to the nearest thousands place.

1. 3,189 → _____

2. 1,492 → _____

Add.

3.
```
    6, 7 8 8
  + 2, 4 6 7
  _____
```

4.
```
    2, 3 5 5
  + 1, 6 7 2
  _____
```

5.
```
   $8. 4 2
  +  3. 2 1
  _____
```

Write the number and say it.

6. 3,000 + 100 + 80 + 8 = _____

Add to find the perimeter.

7.

Subtraction review.

8. 14 – 7 = _____

9. 11 – 6 = _____

10. 7 – 3 = _____

11. 12 – 3 = _____

12. 9 – 5 = _____

13. 8 – 4 = _____

14. Danny picked up five eggs in each hand, and then dropped three eggs. How many whole eggs are left? Be careful!

_____

15. The first time Katherine shot her arrow, it went 296 feet. The second time she shot it, the arrow flew another 316 feet. How many feet did the arrow fly altogether?

_____

16. Tom spent $16 for a shirt, $18 for a pair of pants, and $9 for a tie. How much did he spend on his new clothes?

_____

Round to the nearest ten thousands place.

1. 36,254 → _____

2. 84,531 → _____

Add.

3.   1, 4 7 6
     + 7, 8 1 3

4.   1, 6 2 1
     + 4, 1 5 7

5.
        $7. 1 6
     +   2. 7 9

Write the number and say it.

6. five thousand, four hundred = _____

Add to find the perimeter. Remember that 15' means 15 feet.

7.

```
              15'
        ┌─────────┐
  15'   │         │  15'
        └─────────┘
              15'
```

Subtraction review.

8. 12 – 4 = _____

9. 10 – 3 = _____

10. 7 – 4 = _____

11. 3 – 2 = _____

12. 15 – 6 = _____

13. 9 – 3 = _____

14. Corey has three dimes and four pennies. How much money does he have?

_____

15. Greg has 75 pressed flowers and 108 dried leaves in his collection. How many items is that?

_____

16. Roy counted 1,575 mosquitoes at his picnic. Ron said there were 1,892 mosquitoes at the picnic his family went to. How many mosquitoes in all were going to picnics?

_____

Round to the nearest thousands place.

1. 1,206 → _____   2. 8,738 → _____

Add.

3. $\begin{array}{r} 7,438 \\ + 9,114 \\ \hline \end{array}$ 4. $\begin{array}{r} 6,408 \\ + 4,379 \\ \hline \end{array}$

5. $\begin{array}{r} \$2.56 \\ + 1.24 \\ \hline \end{array}$

Write the number and say it.

6. one hundred thirty-one thousand, five hundred twenty-eight =

_____

Add to find the perimeter.

7.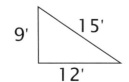

Subtraction review.

8. $13 - 4 =$ _____

9. $11 - 3 =$ _____

10. $14 - 5 =$ _____

11. $10 - 9 =$ _____

12. $6 - 0 =$ _____

13. $8 - 8 =$ _____

14. Twelve people came to Madison's birthday party. Seven of them won prizes. How many did not win prizes?

_____

15. Kitty drove 235 miles yesterday and 256 miles today. Estimate how far she drove, and then find the exact answer.

_____

16. Paula gathered flowers in her garden. She cut 10 roses, 12 lilies, and 18 daisies. How many flowers are in her bouquet?

_____

18A

Round to the nearest hundred and estimate. Then find the exact answer.
The lines are to help you keep your place. The first one is done for you.

1.
$$\begin{array}{r} \overset{1}{4}\,\overset{1}{7}\,6 \\ 3\,3\,5 \\ 2\,2\,5 \\ +\quad 4\,0\,1 \\ \hline 1,4\,3\,7 \end{array}$$

(500)
(300)
(200)
+　(400)
(1,400)

2.
$$\begin{array}{r} 2\,9\,4 \\ 1\,8\,7 \\ 3\,0\,6 \\ +\ 8\,1\,3 \\ \hline \end{array}$$

(　　　)
(　　　)
(　　　)
+ (　　　)
(　　　)

3.
$$\begin{array}{r} 4\,9\,3 \\ 2\,1\,5 \\ 4\,8\,5 \\ 3\,2\,4 \\ +\ 1\,0\,6 \\ \hline \end{array}$$

(　　　)
(　　　)
(　　　)
(　　　)
+ (　　　)
(　　　)

4.*
$$\begin{array}{r} 6\,1\,3 \\ 9\,7 \\ 4\,5\,2 \\ 8\,7\,9 \\ +\quad 3\,0 \\ \hline \end{array}$$

(　　　　)
(1 0 0)
(　　　　)
(　　　　)
+ (　　0)
(　　　　)

*The numbers in a column will not always add to 10. In the first column of this
question, we have 7 + 3 = 10 and 9 + 2 = 11. If necessary, 10 + 11 can be written
to the side and added together before continuing with the rest of the problem.

5.

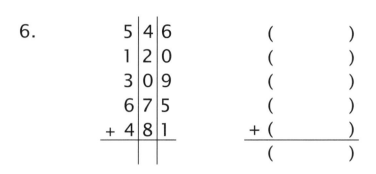

```
    1 1 3        (        )
    2 5 1        (        )
    3 4 5        (        )
    3 5 5        (        )
+   4 2 7     + (        )
             (        )
```

6.

```
    5 4 6        (        )
    1 2 0        (        )
    3 0 9        (        )
    6 7 5        (        )
+   4 8 1     + (        )
             (        )
```

If you wish, turn a sheet of notebook paper sideways and use the lines to keep your numbers lined up.

7. Mrs. Brown fixed up her new house. She spent $127 on paint, $475 for a new chair, and $225 for curtains. First estimate about how much money she spent, and then find the exact answer.

_____

8. Nancy drove to visit her sister. She drove 390 miles the first day, 240 miles the second day, and 152 miles the third day. About how many miles did she drive (estimate)? What is the exact number of miles she drove?

_____

18B

Round to the nearest hundred and estimate. Then find the exact answer. Don't forget to look for 10.

1.
```
  2 6 4        (        )
    8 5        (        )
  6 2 4        (        )
+ 9 4 5      + (        )
             (        )
```

2.
```
  1 7 2        (        )
  2 6 1        (        )
  5 2 7        (        )
+ 4 4 6      + (        )
             (        )
```

3.
```
  9 3 3        (        )
    5 8        (        )
  3 6 1        (        )
  1 5 9        (        )
+ 5 4 2      + (        )
             (        )
```

4.
```
  1 4 2        (        )
  2 0 6        (        )
  8 6 0        (        )
  4 6 2        (        )
+ 5 5 3      + (        )
             (        )
```

5.

```
  3 2 1          (          )
    3 9          (          )
  6 8 6          (          )
  4 5 2          (          )
+ 1 5 2        + (_____)
               (          )
```

6.

```
  2 1 4          (          )
  5 9 6          (          )
  4 7 3          (          )
  5 2 7          (          )
+ 8 0 2        + (_____)
               (          )
```

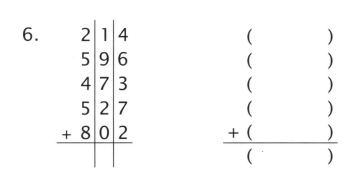

7. In one year, Burton bought 208 gallons of gasoline for his little car, 316 gallons for his big car, and 365 gallons for his pickup truck. How many gallons of gasoline did Burton buy?

_____

8. In Jane's school there are 137 first graders, 122 second graders, 101 third graders, and 150 fourth graders. How many students are in those four grades?

_____

18C

Round to the nearest hundred and estimate. Then find the exact answer. Don't forget to look for 10.

1.
```
    6 4 9          (        )
    5 3 6          (        )
      3 1          (        )
  + 2 2 4      +   (        )
               (        )
```

2.
```
    7 1 4          (        )
    7 4 6          (        )
    4 1 9          (        )
  + 6 5 2      +   (        )
               (        )
```

3.
```
    3 2 8          (          )
    5 3 0          (          )
    3 5 6          (          )
    4 3 2          (          )
  + 4 5 6      +   (          )
               (          )
```

4.
```
    2 1 9          (          )
    8 2 0          (          )
    4 7 9          (          )
    3 8 1          (          )
  + 1 6 1      +   (          )
               (          )
```

5.

```
    5 6 2          (            )
    5 2 0          (            )
      3 8          (            )
    6 5 9          (            )
  + 6 1 3        + (            )
                   (            )
```

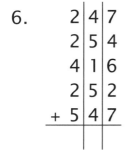
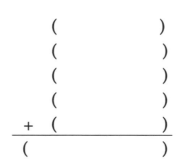

6.

```
    2 4 7          (            )
    2 5 4          (            )
    4 1 6          (            )
    2 5 2          (            )
  + 5 4 7        + (            )
                   (            )
```

7. Julia was bored, so she counted the flakes in a box of raisin bran. She found that there were 553 itty-bitty ones, 334 medium-sized ones, and 129 large ones. How many flakes were in the box of raisin bran?

_____

8. Shirley's collection of soda cans included 302 Coke cans, 148 Pepsi cans, and 447 other cans. How many soda cans did Shirley have?

_____

Add.

1.
```
  2 9 6
  3 0 8
  7 4 2
  8 6 6
+ 3 1 4
```

2.
```
  5 2 1
    5 2
  6 2 4
  5 4 6
+   3 8
```

3.
```
  2 7 3
  9 5 1
  5 5 0
  3 3 9
+ 2 8 2
```

Write the number and say it.

4. 7,000 + 800 + 20 + 1 = _____

5. three hundred fifty thousand = _____

Subtraction review.

6. $\begin{array}{r} 2 \\ -\ 0 \\ \hline \end{array}$ 7. $\begin{array}{r} 4 \\ -\ 1 \\ \hline \end{array}$

8. $\begin{array}{r} 6 \\ -\ 2 \\ \hline \end{array}$ 9. $\begin{array}{r} 5 \\ -\ 5 \\ \hline \end{array}$

Challenger: See if you can write the correct sign in each blank. The first two are done for you.

10. 5 $-$ 3 = 2   11. 6 $+$ 1 = 7

12. 10 __ 8 = 2   13. 2 __ 5 = 7

Skip count and write the numbers.

14. _____, 4, _____, 8, _____, _____, _____, _____, _____, _____

15. At Rachel's wedding, 51 guests were very early and 29 guests were just a little early. The other 19 guests were just in time. How many guests did she have?

_____

16. As Tess walked through the meadow, she noticed that there were 68 red flowers, 89 blue flowers, and 132 purple flowers. How many flowers did she count?

_____

Add.

1.
```
  4 5 7
  5 6 1
  4 5 1
  6 3 1
+   2 9
```

2.
```
  8 7 3
  2 6 5
  3 1 4
  2 4 7
+ 9 3 6
```

3.
```
  8 0 4
  2 4 6
  5 3 1
  3 8 2
+ 6 3 9
```

Write the number and say it.

4. 80,000 + 1,000 + 200 + 40 + 6 = _____

5. six hundred fifty-two thousand, six hundred ninety-three =

_____

Subtraction review.

6. 1 2
    – 9

7. 1 5
    – 9

8. 1 1
    – 9

9. 1 7
    – 9

See if you can write the correct sign in each blank.

10. 9 __ 4 = 13

11. 18 __ 9 = 9

12. 14 __ 9 = 5

13. 9 __ 7 = 16

Skip count and write the numbers.

14. ____, ____, 15, 20, ____, ____, ____, ____, ____, ____

15. Madison's Christmas tree was six feet tall. How many inches tall was her tree?

     _____

16. Bailey found three nickels and four dimes. How much money did she find?

     _____

Add.

1.
```
  7|4|1
  4|0|5
  4|0|1
  5|8|6
+ 3|6|4
```

2.
```
  9|5|2
  2|3|7
  1|7|1
  6|0|3
+ 4|5|9
```

3.
```
  6|3|1
  8|3|4
  8|4|1
  2|7|2
+ 2|3|4
```

Write the number and say it.

4. $600{,}000 + 30{,}000 + 7{,}000 + 500 + 30 + 1 =$ _____

5. forty-five thousand, seven hundred twenty-seven =

_____

Subtraction review.

6. 1 1
    – 8

7. 1 4
    – 8

8. 1 7
    – 8

9. 1 3
    – 8

See if you can write the correct sign in each blank.

10. 15 __ 8 = 7

11. 12 __ 8 = 4

12. 8 __ 8 = 16

13. 9 __ 8 = 1

Skip count and write the numbers.

14. ____, 20, ____, ____, 50, ____, ____, ____, ____, ____

15. Liane read three books. The first had 87 pages, the next had 48 pages, and the last had 211 pages. How many pages did Liane read?

    _____

16. Caleb hopes to get his license when he is 16 years old. If he is 7 years old now, how many years does he have to wait?

    _____

19A

Round to the nearest thousand and estimate the answer. Then find the exact answer. Don't forget to look for 10. The first one is done for you.

1.
```
  1  1
 1, 5 8 2
 3, 6 2 4
+   4 1 3
 5, 6 1 9
```
$$(2, 0\ 0\ 0)$$
$$(4, 0\ 0\ 0)$$
$$+(\ \ \ 0\ 0\ 0)$$
$$(6, 0\ 0\ 0)$$

2.
```
  7, 1 3 2
  5, 3 3 3
+ 1, 1 8 6
```
$$(\qquad\qquad)$$
$$(\qquad\qquad)$$
$$+\ (\qquad\qquad)$$
$$(\qquad\qquad)$$

3.
```
  2, 8 5 2
  4, 2 6 3
+ 3, 1 4 9
```
$$(\qquad\qquad)$$
$$(\qquad\qquad)$$
$$+\ (\qquad\qquad)$$
$$(\qquad\qquad)$$

4.
```
  6, 7 3 2
  3, 1 5 2
+ 7, 3 2 1
```
$$(\qquad\qquad)$$
$$(\qquad\qquad)$$
$$+\ (\qquad\qquad)$$
$$(\qquad\qquad)$$

5.

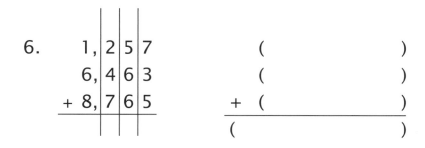

	5,	2	3	2
	7,	1	1	1
+	3,	7	6	5

(   )
(   )
+ (   )
(   )

6.

	1,	2	5	7
	6,	4	6	3
+	8,	7	6	5

(   )
(   )
+ (   )
(   )

You may turn a sheet of notebook paper sideways and use the lines to keep your columns lined up.

7. Claire traveled from Boston to Atlanta (1,083 miles), then to Tampa (482 miles), then to Detroit (1,216 miles) and then back to Boston (741 miles). How many miles did she travel in all?

_____

8. For Claire's next adventure she flew to San Francisco (3,187 miles), then on to Los Angeles (408 miles), and from there to Dallas (1,443 miles). After that she flew home to Boston (1,837 miles). How many miles did she fly altogether?

_____

19B

Round to the nearest thousand first and estimate the answer. Then find the exact answer. Don't forget to look for 10.

1.
```
   2,8 1 7        (            )
   9,2 3 6        (            )
 + 3,6 8 0      + (            )
                  (            )
```

2.
```
   5,7 4 0        (            )
   4,2 2 1        (            )
 + 3,3 2 1      + (            )
                  (            )
```

3.
```
   3,2 1 3        (            )
   1,3 5 7        (            )
 + 2,7 9 8      + (            )
                  (            )
```

4.
```
   1,4 7 6        (            )
     7 4 6        (            )
 + 9,8 1 3      + (            )
                  (            )
```

5.
```
    2, 7 4 1        (            )
    4, 3 7 4        (            )
  + 3, 3 5 4      + (            )
                    (            )
```

6.
```
    4, 1 2 3        (            )
    7, 4 9 1        (            )
  +    4 8 6      + (            )
                    (            )
```

7. The fisherman caught 2,365 fish the first day, 1,295 fish the second day, and 3,116 the third day. How many fish did he catch in all?

   _____

8. In Pineville 7,851 students go to school and 3,895 are homeschooled. How many students live in Pineville altogether?

   _____

Round to the nearest thousand first and estimate the answer. Then find the exact answer. Don't forget to look for 10.

1.
```
   3,6 9 5
   3,1 7 5
 + 2,1 4 1
```
(   )
(   )
+ (   )
(   )

2.
```
   1,4 6 8
   6,0 1 2
 + 5,2 8 0
```
(   )
(   )
+ (   )
(   )

3.
```
   2,9 4 0
   4,2 7 8
 + 1,9 6 3
```
(   )
(   )
+ (   )
(   )

4.
```
   1,0 8 6
   3,6 0 8
 + 2,6 5 7
```
(   )
(   )
+ (   )
(   )

5.

```
    1, 3 4 0          (              )
    1, 7 6 5          (              )
  + 8, 0 4 1        + (              )
                      (              )
```

6.

```
    4, 7 2 9          (              )
    7, 7 6 1          (              )
  +       3 5       + (              )
                      (              )
```

7. Smithville has 8,716 people living in it, and Jonesville has 6,658 people. How many people live in the two towns combined?

_____

8. Mr. White has a furniture store. On Monday he had three customers who spent $2,345, $5,634, and $1,954. How much did they spend in all?

_____

Add.

1.
$$
\begin{array}{r}
2,475 \\
1,890 \\
+376 \\
\hline
\end{array}
$$

2.
$$
\begin{array}{r}
7,513 \\
9,025 \\
+\,3,254 \\
\hline
\end{array}
$$

3.
$$
\begin{array}{r}
3,189 \\
1,422 \\
+\,2,468 \\
\hline
\end{array}
$$

Subtraction review. The next lesson will begin multiple-digit subtraction!

4.
$$
\begin{array}{r}
8 \\
-\,4 \\
\hline
\end{array}
$$

5.
$$
\begin{array}{r}
10 \\
-\,6 \\
\hline
\end{array}
$$

6.
$$
\begin{array}{r}
14 \\
-\,7 \\
\hline
\end{array}
$$

7.
$$
\begin{array}{r}
10 \\
-\,9 \\
\hline
\end{array}
$$

8.
$$
\begin{array}{r}
6 \\
-\,3 \\
\hline
\end{array}
$$

9.
$$
\begin{array}{r}
10 \\
-\,3 \\
\hline
\end{array}
$$

10.
$$\begin{array}{r} 4 \\ -\ 2 \\ \hline \end{array}$$

11.
$$\begin{array}{r} 1\ 0 \\ -\ 5 \\ \hline \end{array}$$

Compare, and then fill in the oval with <, >, or =.

12. $12 - 6 \bigcirc 2 + 2$

13. $10 + 4 \bigcirc 10 - 4$

14. $10 - 7 \bigcirc 10 - 3$

15. Jason earned $35 on Monday, $42 on Tuesday, $33 on Wednesday, and $45 on Thursday. How much did he earn in all?

———————————————

16. Hannah needs $17 to buy a gift for her mother. She already has $8. How much more money does she need?

———————————————

17. Seth is raising earthworms to sell. One box has 3,645 earthworms, one has 4,782, and the last has 5,641. How many earthworms does Seth have?

———————————————

18. Alexis drew a triangle. Each side was five inches long. What was the perimeter of her triangle?

———————————————

Add.

1.
```
  2,3 8 4
  4,1 2 3
+ 6,3 3 5
```

2.
```
  3,5 9 1
  2,3 6 7
+ 8,4 5 9
```

3.
```
  2,3 6 8
  4,1 5 2
+ 6,3 1 4
```

Subtraction review. The next lesson will begin multiple-digit subtraction!

4.
```
   9
 - 5
```

5.
```
   2
 - 1
```

6.
```
   9
 - 6
```

7.
```
  1 0
 - 2
```

8.
```
   9
 - 3
```

9.
```
  1 6
 - 9
```

10. 9
    − 4
    ─────

11. 9
    − 7
    ─────

Compare, and then fill in the oval with <, >, or =.

12. 12 − 3 ◯ 12 + 3   13. 16 − 7 ◯ 14 − 5

14. 6 − 4 ◯ 9 − 1

15. Forrest bought a birthday card that cost $2.98 and a balloon that cost $3.75. How much did he spend in all?

    _____

16. Paula found seven dimes, and then lost five of them. How many cents does she have left?

    _____

17. A builder is ordering cement. For different jobs he needs 20 tons, 18 tons, 10 tons, and 35 tons. How many tons of cement should he order?

    _____

18. Mr. Brown's vegetable garden is shaped like a square with every side 35 feet long. How many feet of fence does he need to go all around his garden?

    _____

Add.

1.
$$\begin{array}{r} 8,482 \\ 4,621 \\ +\ 5,351 \\ \hline \end{array}$$

2.
$$\begin{array}{r} 1,264 \\ 7,632 \\ +\ 1,953 \\ \hline \end{array}$$

3.
$$\begin{array}{r} 5,148 \\ 2,633 \\ +\ 4,186 \\ \hline \end{array}$$

Subtraction review. Be sure you know your subtraction facts before going to the next lesson.

4.
$$\begin{array}{r} 7 \\ -\ 3 \\ \hline \end{array}$$

5.
$$\begin{array}{r} 11 \\ -\ 6 \\ \hline \end{array}$$

6.
$$\begin{array}{r} 7 \\ -\ 4 \\ \hline \end{array}$$

7.
$$\begin{array}{r} 13 \\ -\ 5 \\ \hline \end{array}$$

8.
$$\begin{array}{r} 8 \\ -\ 3 \\ \hline \end{array}$$

9.
$$\begin{array}{r} 9 \\ -\ 2 \\ \hline \end{array}$$

10. 8
    - 5

11. 1 1
    – 4

Compare, and then fill in the oval with <, >, or =.

12. 8 – 6 $\bigcirc$ 8 – 7

13. 12 – 7 $\bigcirc$ 12 + 7

14. 14 – 6 $\bigcirc$ 4 + 4

15. A truck driver drove the following numbers of miles on different days: 495, 382, 516, and 402. How many miles did he drive in all?

_____

16. Dave's yard is a rectangle. Two sides are 25 feet and two sides are 35 feet. What is the perimeter of his yard?

_____

17. Aunt Dot knitted mittens for six people. If each person has two hands, how many mittens did she need to make? (skip count)

_____

18. Four of the mittens Aunt Dot knitted were red and the rest were blue. Use your answer for number 17 and find out how many mittens were blue.

_____

Subtract and check by adding. The first one is done for you.
(Choose the way of checking you like best. You don't need to
check the problem twice!)

1.
```
    7 2        check
  – 3 1          3 1
  -----        + 4 1
    4 1        -----
    7 2          7 2
```

2.
```
    6 0
  – 4 0
```

3.
```
    9 4
  – 5 1
```

4.
```
    5 3
  – 4 2
```

5.
```
    4 0
  – 3 0
```

6.
```
    4 5 9
  – 3 1 2
```

7.
```
    4 4
  – 2 0
```

8.
```
    9 2 4
  –   1 3
```

9.
$$\begin{array}{r} 5\ 0\ 6 \\ -\ 3\ 0\ 2 \\ \hline \end{array}$$

10.
$$\begin{array}{r} 2\ 5 \\ -\ 2\ 1 \\ \hline \end{array}$$

11.
$$\begin{array}{r} 8\ 4\ 1 \\ -\ 6\ 2\ 0 \\ \hline \end{array}$$

12.
$$\begin{array}{r} 9\ 9\ 9 \\ -\ 1\ 2\ 3 \\ \hline \end{array}$$

13. There are 32 students in Mrs. Martin's class. Ten students are out with the flu today. How many are in the class today?

_____

14. Emaleigh raised rabbits. She had 44 rabbits, but someone left the cage doors open by accident, and 11 rabbits ran away. How many rabbits are left?

_____

Subtract and check by adding.

1.
```
    3 5
  - 2 4
```

2.
```
    2 6
  - 1 3
```

3.
```
    5 0
  - 2 0
```

4.
```
    8 3
  - 1 2
```

5.
```
    4 9
  - 4 7
```

6.
```
    9 8 9
  - 4 3 2
```

7.
```
    4 6
  - 3 0
```

8.
```
  5 5 4
  - 2 1
```

9.
```
  3 0 0
- 1 0 0
```

10.
```
    6 2
 - 3 0
```

11.
```
  4 3 8
- 2 1 4
```

12.
```
    3 9 7
 - 1 7 5
```

13. Daniel collects baseball cards. He had 68 cards, but he traded 25 of them for some special marbles. How many baseball cards did he have left?

_____

14. A truck had 48 gallons of gasoline in the tank when it started its trip. If 23 gallons were left at the end of the trip, how many gallons were used?

_____

Subtract and check by adding.

1.
```
   6 5
 - 3 2
```

2.
```
   1 7
 - 1 7
```

3.
```
   5 2
 - 2 1
```

4.
```
   2 0
 - 1 0
```

5.
```
   7 5
 - 3 2
```

6.
```
   1 8 8
 -   2 3
```

7.
```
   6 9
 - 3 1
```

8.
```
   5 6 1
 - 2 6 0
```

9.
```
   6 4 5
 - 4 3 5
```

10.
```
    8 5
  - 4 4
```

11.
```
   2 2 5
 - 1 0 0
```

12.
```
   5 3 8
 - 4 2 1
```

13. Twenty-nine children at summer camp got prize ribbons. Thirteen children lost their ribbons. How many still had ribbons to take home?

   _____

14. Chad had $49. He spent $21 on a gift for his sister. How much money does Chad have left?

   _____

Subtract and check by adding.

1.　　77
　　− 1 1

2.　　5 0
　　− 4 0

3.　　3 9
　　− 2 7

4.　　6 9 3
　　− 3 6 1

5.　　3 0 0
　　− 1 0 0

6.　　1 6 3
　　−　5 1

Add.

7.　2,1 7 4
　　7,4 1 8
　+ 3,7 9 1

　　(　　　　)
　　(　　　　)
　+ (　　　　)
　　(　　　　)

8.　2,5 6 4
　　6,4 0 8
　+ 1,2 4 3

　　(　　　　)
　　(　　　　)
　+ (　　　　)
　　(　　　　)

9.
```
    3 7 9        (          )
    5 1 1        (          )
    3 3 3        (          )
  + 4 6 8      + (_____)
  ─────────      (          )
```

Write the number and say it.

10. $100,000 + 20,000 + 4,000 + 900 + 70 + 1 =$ _____

Skip count and write the numbers.

11. _____, _____, 6, _____, 10, _____, _____, _____, _____, _____

12. _____, 10, _____, _____, 25, _____, _____, _____, _____, _____

13. _____, _____, 30, _____, _____, _____, _____, _____, _____, 100

14. Carolyn had 25¢ in her pocket. If she spent 14¢, how much did she have left?

_____

15. During their vacation, Jason's family traveled 103 miles by car, 3,521 miles by airplane, and 58 miles by train. How many miles did they travel in all? Be sure you have place values in the correct column when you set up the problem.

_____

Subtract and check by adding.

1.
```
    3 5
  - 2 2
```

2.
```
    4 9
  - 3 1
```

3.
```
    2 8
  -   5
```

4.
```
    6 3 3
  - 5 1 0
```

5.
```
    7 9 0
  - 6 9 0
```

6.
```
    5 6 1
  - 5 5 0
```

Add.

7.
```
    1,8 9 0          (          )
    3,6 7 2          (          )
  + 7,2 5 4      + (            )
                    (            )
```

8.
```
    7,1 9 3          (          )
    4,6 8 5          (          )
  + 1,4 9 2      + (            )
                    (            )
```

9.
$$
\begin{array}{r}
9|2|2 \\
6|7|8 \\
2|5|3 \\
+\ 1|1|2 \\
\hline
\end{array}
$$

```
(            )
(            )
(            )
+  (_____)
(            )
```

Write the number and say it.

10. two hundred five thousand, nine hundred eighteen =

   _____

Skip count and write the numbers.

11. 5, _____, _____, _____, _____, _____, _____, _____, _____, _____

12. _____, 4, _____, _____, _____, 12, _____, _____, _____, _____

13. _____, _____, _____, 40, _____, _____, _____, _____, 90, _____

14. Sandi had 49 pennies. If she bought a candy bar that cost 23 cents, how much money did she have left over?

   _____

15. A factory bought 285 boxes of apples, 683 boxes of peaches, and 491 boxes of pears to make juice. How many boxes of fruit were bought in all?

   _____

Subtract and check by adding.

1.
```
    9 0
  - 7 0
```

2.
```
    5 1
  - 1 0
```

3.
```
    7 4
  - 3 1
```

4.
```
    1 3 9
  - 1 2 5
```

5.
```
    9 9 9
  - 3 4 7
```

6.
```
  1 6 7
  - 2 4
```

Add.

7.
```
    2,4 6 7        (          )
    9,2 2 1        (          )
  + 6,7 8 8      + (          )
                   (          )
```

8.
```
    4,2 5 3        (          )
    4,1 1 2        (          )
  + 1,2 5 5      + (          )
                   (          )
```

9.
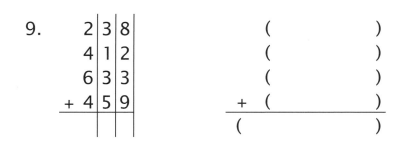

Write the number and say it.

10.   20,000 + 3,000 + 600 + 80 + 4 = _____

Skip count and write the numbers.

11. ____, 20, ____, ____, ____, ____, 70, ____, ____, ____

12. ____, ____, 15, ____, ____, ____, 35, ____, ____, ____

13. ____, ____, ____, 8, ____, ____, 14, ____, ____, ____

14. A yardstick is three feet long. How many inches long is it?

_____

15. Jacki counted 67 cars on the way to Grandma's house. That is 13 more than Vicki saw. Subtract to find how many cars Vicki saw.

_____

21A

Write how many minutes are shown by each clock. The first one is done for you. (There is a clock template at the end of this book.)

1.

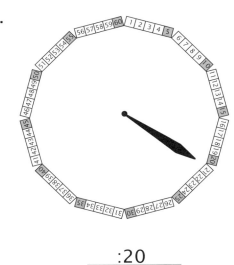

_____:20_____

2.

_____

3.

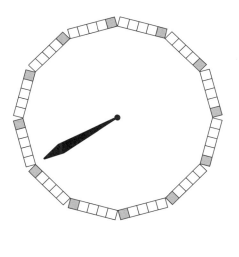

_____

4.

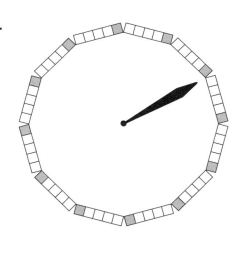

_____

5.

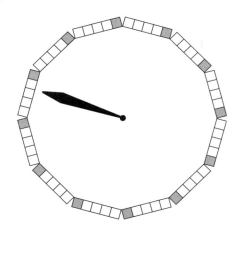

_____

6.

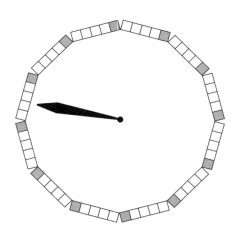

_____

Write how many minutes are shown by each clock.

1.

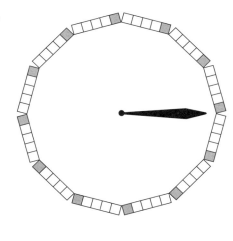

_____

2.

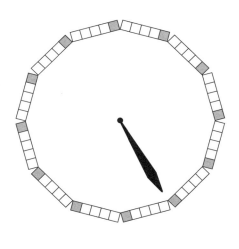

_____

3.

_____

4.

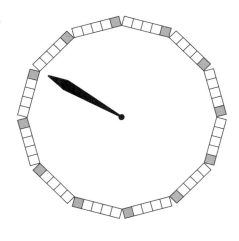

_____

5.

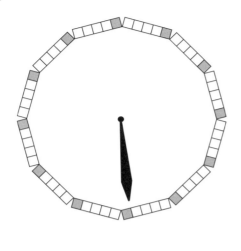

_____

6.

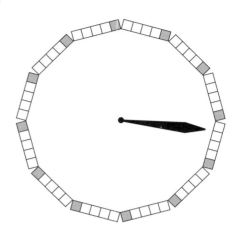

_____

Write how many minutes are shown by each clock.

1.

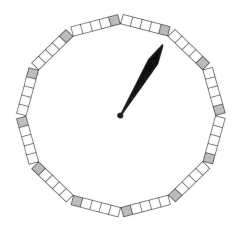

_____

2.

_____

3.

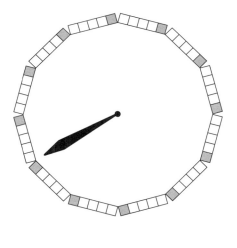

_____

4.

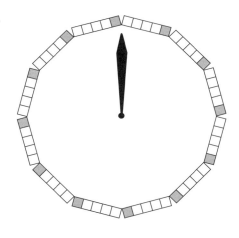

_____

5.

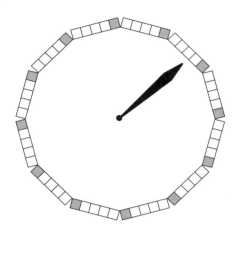

_____

6.

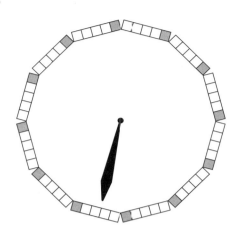

_____

21D

Write how many minutes are shown by each clock.

1.

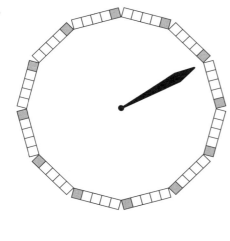

_____

2.

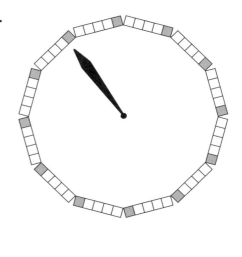

_____

Subtract and check.

3.   9 9
  − 8 4

4.   1 3 8
   − 3 4

5.   2 7 9
  − 1 6 4

Add.

6. $3. 2 7
  + 4. 1 6

7.   2,3 8 4
 + 6,3 3 5

8.   3 6 7
   5 9 1
  + 4 1 9

9. In the spring the bugs came out to play. In John's back yard there were 4,568 gnats, 3,851 mosquitoes, and 2,001 dragonflies. How many insects were there altogether?

_____

10. Keith is buying fencing to make a square pen for the family's pet goat. The pen will be 10 feet on each side. The gate will be 3 feet wide. How many feet of fence should Keith buy? (Be careful, this is a two-step problem.)

_____

11. Rose is 25 and Karen is 14. What is the difference in their ages?

_____

12. Dan made 19 points playing a game, and Angie made 7 points. How many more points did Dan make?

_____

Write how many minutes are shown by each clock.

1.

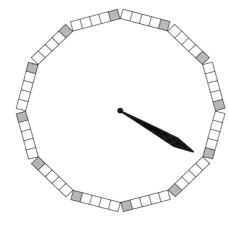

2.

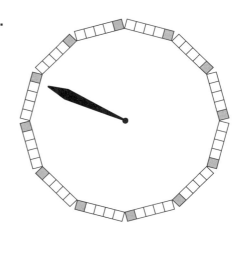

_____   _____

Subtract and check.

3.   8 3
     – 5 2

4.   8 7 3
     – 6 1

5.   3 6 2
     – 1 5 1

Add.

6. $9. 2 8
     + 2. 1 7

7.   4, 1 2 3
     + 7, 4 6 0

8.   4 5 9
      3 6 7
     + 5 9 1

9. Lillian is 28 years older than Pam. If Pam is 32, how old is Lillian?

_____

10. Russ is 8 years younger than Jim. If Jim is 29, how old is Russ?

_____

11. Greg walked all around the outside walls of the house. His house is shaped like a rectangle. It is 40 feet long and 35 feet wide. How many feet did Greg walk?

_____

12. Mom made 144 chocolate chip cookies and 120 raisin cookies. After the picnic, only 52 cookies were left over. How many cookies were eaten at the picnic?

_____

Write how many minutes are shown by each clock.

1.

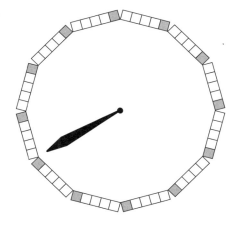

_____

2.

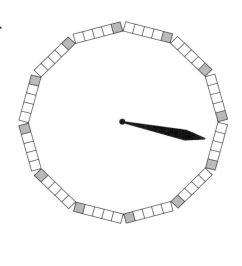

_____

Subtract and check.

3.   1 9
     –   7
  _____

4.   8 3 9
   – 1 6
  _____

5.   6 0 4
     – 3 0 2
  _____

Add.

6. $5. 3 6
    + 1. 5 1
  _____

7.   8,4 8 2
   + 4,6 2 1
  _____

8.   3 4 3
       2 4 6
     + 1 1 2
  _____

9. Shirley found four dimes and three nickels, and then she lost 25¢. How much money does she have left?

   _____

10. Simon drew a triangle that was six inches long on each side. What was the perimeter of the triangle?

    _____

11. April was very bored, so she started counting leaves on the bushes in the back yard. She counted the following numbers of leaves: 969, 1,345, 5,002, and 2,061. How many leaves did she count in all?

    _____

12. Clyde needs $25 to buy a gift for his mom. If he has $10, how many more dollars does he need?

    _____

Subtract using regrouping. Check by adding using the way you like best.
The first one is done for you.

1.
$$
\begin{array}{r}
\overset{3}{\cancel{4}}\overset{1}{2} \\
-\ 1\ 8 \\
\hline
2\ 4 \\
\hline
4\ 2
\end{array}
$$

check
$$
\begin{array}{r}
1\ 8 \\
+\ 2\ 4 \\
\hline
4\ 2
\end{array}
$$

2.
$$
\begin{array}{r}
6\ 1 \\
-\ 2\ 7 \\
\hline
\end{array}
$$

3.
$$
\begin{array}{r}
2\ 2 \\
-\ 1\ 3 \\
\hline
\end{array}
$$

4.
$$
\begin{array}{r}
2\ 3 \\
-\ 1\ 9 \\
\hline
\end{array}
$$

5.
$$
\begin{array}{r}
5\ 5 \\
-\ 2\ 6 \\
\hline
\end{array}
$$

6.
$$
\begin{array}{r}
3\ 2 \\
-\ 1\ 4 \\
\hline
\end{array}
$$

7.
$$
\begin{array}{r}
7\ 3 \\
-\ 3\ 4 \\
\hline
\end{array}
$$

8.
$$
\begin{array}{r}
6\ 8 \\
-\ 2\ 9 \\
\hline
\end{array}
$$

9.
$$
\begin{array}{r}
6\ 3 \\
-\ 4\ 9 \\
\hline
\end{array}
$$

10. Kelsey had 75¢. She paid 57¢ for a yo-yo. How much money did she have left?

_____

11. Briley made 43 fancy barrettes to sell at a craft fair. If she sold 28 of them, how many did she have left?

_____

12. An elephant ate 65 peanuts on Tuesday and 72 peanuts on Thursday. How many more peanuts did it eat on Thursday?

_____

22B

Subtract using regrouping. Check by adding.

1.
$$\begin{array}{r} 5\ 7 \\ -\ 2\ 9 \\ \hline \end{array}$$

2.
$$\begin{array}{r} 3\ 0 \\ -\ 1\ 8 \\ \hline \end{array}$$

3.
$$\begin{array}{r} 6\ 5 \\ -\ 4\ 7 \\ \hline \end{array}$$

4.
$$\begin{array}{r} 5\ 2 \\ -\ 1\ 4 \\ \hline \end{array}$$

5.
$$\begin{array}{r} 8\ 5 \\ -\ 4\ 9 \\ \hline \end{array}$$

6.
$$\begin{array}{r} 7\ 2 \\ -\ 2\ 7 \\ \hline \end{array}$$

7.
$$\begin{array}{r} 5\ 3 \\ -\ 1\ 8 \\ \hline \end{array}$$

8.
$$\begin{array}{r} 3\ 1 \\ -\ 1\ 6 \\ \hline \end{array}$$

9.
$$\begin{array}{r} 8\ 2 \\ -\ 5\ 3 \\ \hline \end{array}$$

10. Wayne's book has 43 pages. If he has read 24 pages, how many does he have left to read?

    _____

11. Stan tossed the basketball at the hoop 61 times. If he missed 36 times, how many times did the ball go through the hoop?

    _____

12. Kitty has 50 cents. How much would she have left if she bought a 23-cent candy bar?

    _____

22C

Subtract using regrouping. Check by adding.

1.
```
  5 2
- 2 5
```

2.
```
  7 1
-   3
```

3.
```
  3 4
- 1 5
```

4.
```
  8 7
-   8
```

5.
```
  6 2
- 2 7
```

6.
```
  2 3
- 1 4
```

7.
```
  4 5
- 2 6
```

8.
```
  3 1
- 2 9
```

9.
```
  7 2
- 3 8
```

10. Hannah had 65 pennies, but she lost six of them. How many pennies does she have left?

   _____

11. Sue is 48 and Richard is 29. What is the difference in their ages?

   _____

12. Stacia bought apples to make pies. There were 63 apples in the basket, but 24 of them were spoiled. How many apples does Stacia have to use?

   _____

Subtract, using regrouping when needed. Check by adding.

1.　　7 1
　　− 4 3

2.　　9 8
　　− 8 9

3.　　3 5
　　− 1 7

4.　　8 5
　　− 4 5

5.　　4 2 9
　　− 3 1 1

6.　1 4 8
　　− 2 6

Add.

7.　　$7. 3 3
　　+　1. 8 3

8.　　5,2 6 3
　　+ 7,5 5 4

9.　　8 9 2
　　　4 1 3
　　+ 4 7 6

Write how many minutes are shown by each clock.

10.

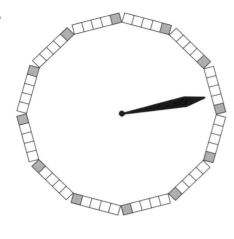

_____

11.

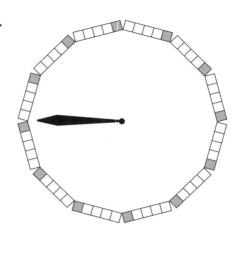

_____

12. Anthony collected 16 red marbles, 21 green marbles, and 14 blue marbles. He gave 25 of his marbles to Jack. How many marbles did Anthony have left?

_____

22E

Subtract, using regrouping when needed. Check by adding.

1.
```
   9 2
 - 7 8
```

2.
```
   5 3
 - 4 5
```

3.
```
   2 3
 - 1 4
```

4.
```
   7 5
 - 5 0
```

5.
```
  7 4 3
 -  1 2
```

6.
```
  5 1 4
 -    2
```

Add.

7.
```
  $3. 5 5
 +  2. 5 2
```

8.
```
   9, 4 3 5
 + 4, 2 5 2
```

9.
```
    5 7 3
    8 8 2
    2 3 5
 +  7 6 2
```

Write how many minutes are shown by each clock.

10.

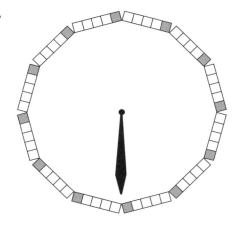

_____

11.

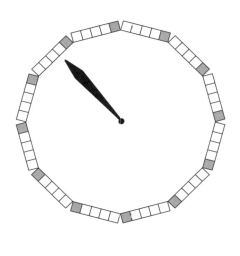

_____

12. Esther read 19 pages in her book yesterday and 21 pages today. If there are 52 pages in all in her book, how many does she have left to read?

_____

22F

Subtract, using regrouping when needed. Check by adding.

1.　　44
　　− 28

2.　　93
　　− 16

3.　　84
　　− 27

4.　　17
　　− 10

5.　　913
　　− 112

6.　　307
　　− 205

Add.

7.　$ 6. 2 4
　+ 5. 4 8

8.　　3,5 4 8
　　　7,6 2 4
　+ 3,6 4 2

9.　　5 7 3
　　　8 8 2
　+ 7 6 2

Write how many minutes are shown by each clock.

10.

_____

11.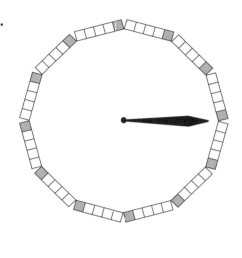

_____

12. Ranger Bill counted 134 deer, 48 elk, and 2 mountain lions. Ranger Tom counted 100 deer and 22 elk. How many more animals did Ranger Bill count?

_____

What is the hour? The first one is done for you.

1. ___9:00___

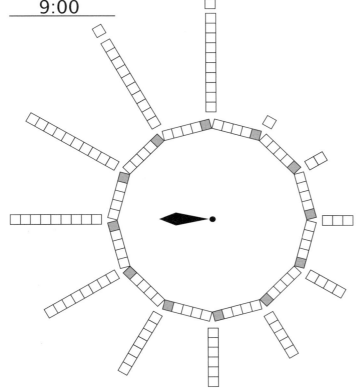

2. _____

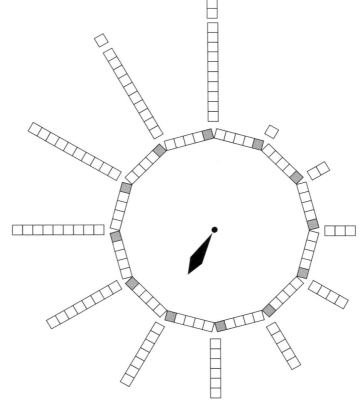

3. _____

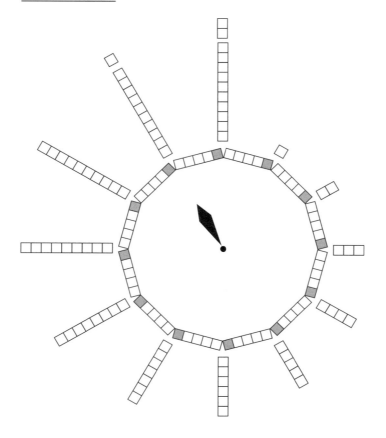

4. _____

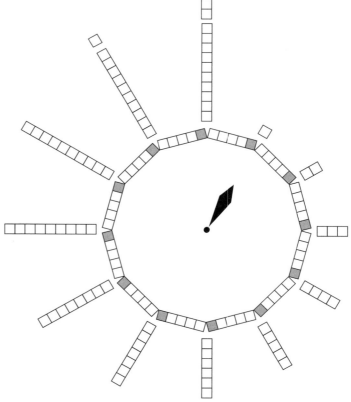

What is the hour?

1. _____

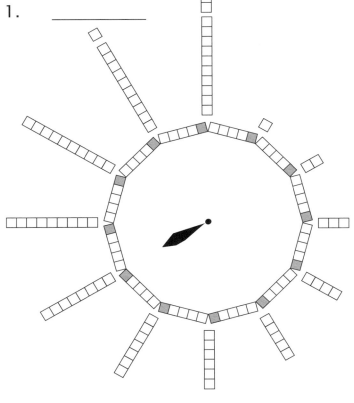

2. _____

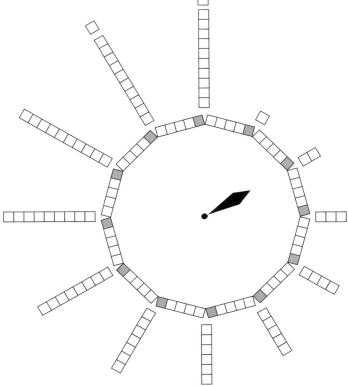

3. _____

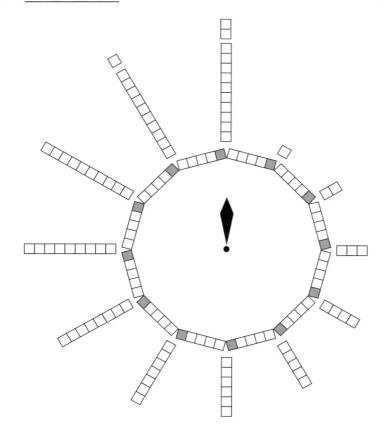

4. _____

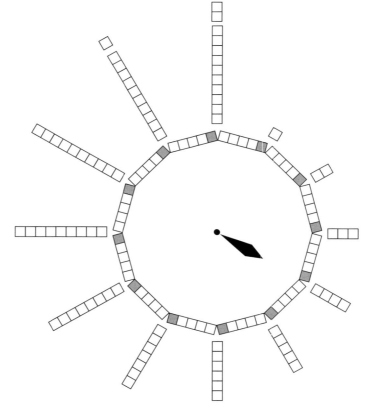

Give the time with hours and minutes. The first one is done for you.

1. 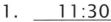 11:30

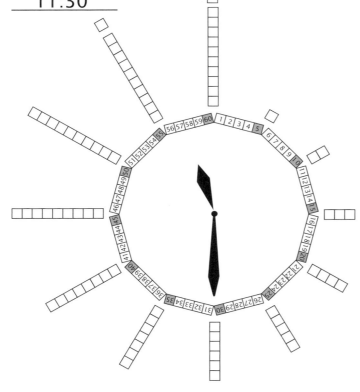

2. _____

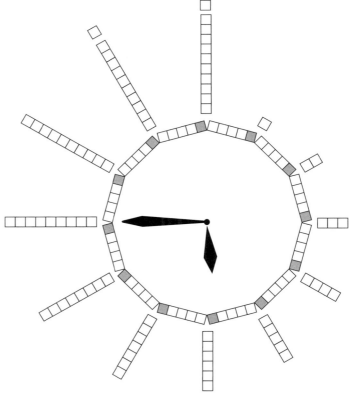

3. _____

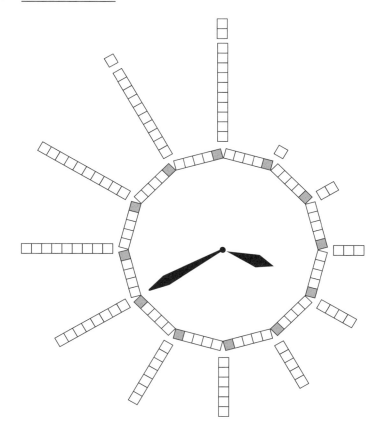

4. _____

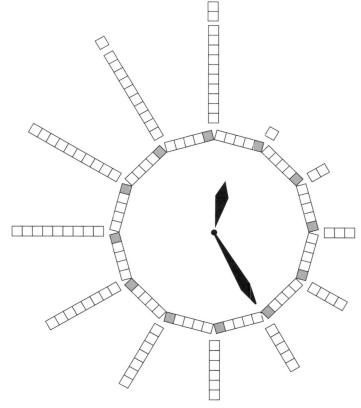

23D

Give the time with hours and minutes.

1.

_____

2.

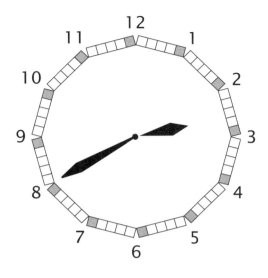

_____

3.

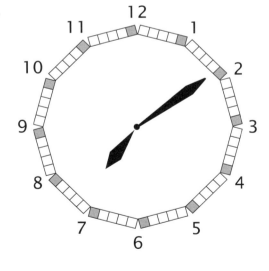

_____

4.

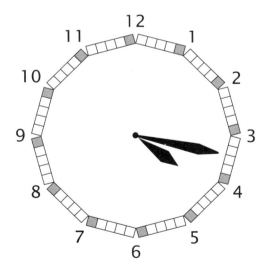

_____

Subtract and check.

5.
```
   4 0
 - 3 8
```

6.
```
   7 6
 - 1 8
```

7.
```
   6 2
 - 5 7
```

8.
```
   7 3
 - 1 4
```

9. Sonia has four dimes, seven nickels, and eight pennies. How much money does she have?

   _____

10. Each side of a square field is 1,342 feet long. What is the perimeter of the field?

   _____

Give the time with hours and minutes.

1.

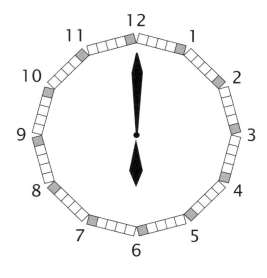

_____

2.

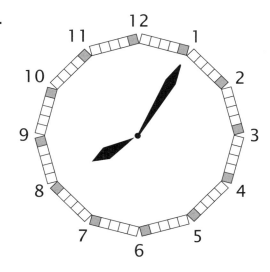

_____

3.

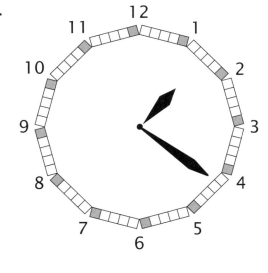

_____

4.

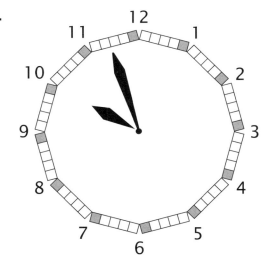

_____

Subtract and check.

5.
$$\begin{array}{r} 7\ 3 \\ -\ 1\ 5 \\ \hline \end{array}$$

6.
$$\begin{array}{r} 8\ 8 \\ -\ 4\ 4 \\ \hline \end{array}$$

7.
$$\begin{array}{r} 7\ 6 \\ -\ 2\ 8 \\ \hline \end{array}$$

8.
$$\begin{array}{r} 9\ 2\ 8 \\ -\ 6\ 0\ 5 \\ \hline \end{array}$$

9. Oliver is six feet tall. How many inches tall is he?

_____

10. Joseph had 250 toy soldiers. He got 75 more from Johnny for his birthday. How many toy soldiers does Joseph have now?

_____

Give the time with hours and minutes.

1.

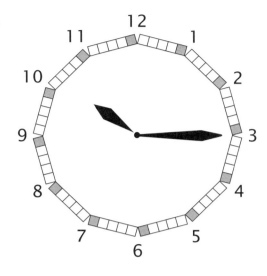

_____

2.

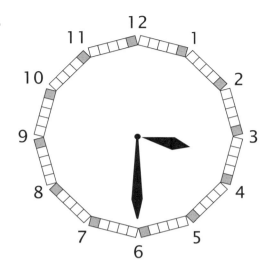

_____

3.

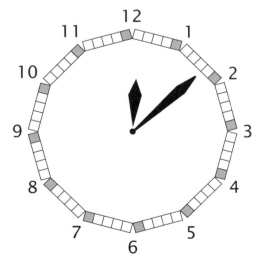

_____

4.

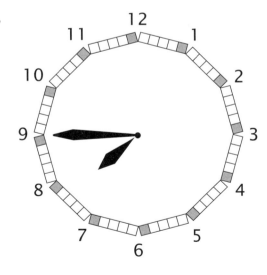

_____

Subtract and check.

5.   5 4
     – 4 5

6.   3 9
     – 2 0

7.   4 0
     – 3 5

8.   7 7 9
     – 3 1 4

9. Kelly bought a salad for $4.52 and a sandwich for $2.91. How much did Kelly spend in all?

   _____

10. Thirty-five birds landed in Sandra's back yard, then twenty-six of them flew away. How many birds were left?

    _____

24A

Subtract using regrouping. Check by adding. The first two are done for you.

1.
$$\begin{array}{r} \overset{1}{\cancel{2}}\overset{9}{\cancel{0}}\overset{1}{0} \\ - \quad 2\ 5 \\ \hline 1\ 7\ 5 \\ 2\ 0\ 0 \end{array}$$

check
$$\begin{array}{r} 2\ 5 \\ +1\ 7\ 5 \\ \hline 2\ 0\ 0 \end{array}$$

2.
$$\begin{array}{r} 8\ \overset{2}{\cancel{3}}\overset{1}{5} \\ -2\ 2\ 8 \\ \hline 6\ 0\ 7 \\ 8\ 3\ 5 \end{array}$$

check
$$\begin{array}{r} 2\ 2\ 8 \\ +6\ 0\ 7 \\ \hline 8\ 3\ 5 \end{array}$$

3.
$$\begin{array}{r} 2\ 2\ 3 \\ -\ 8\ 7 \\ \hline \end{array}$$

4.
$$\begin{array}{r} 7\ 3\ 4 \\ -\ 3\ 6 \\ \hline \end{array}$$

5.
$$\begin{array}{r} 5\ 2\ 3 \\ -1\ 3\ 8 \\ \hline \end{array}$$

6.
$$\begin{array}{r} 4\ 0\ 0 \\ -3\ 9\ 9 \\ \hline \end{array}$$

7.
$$\begin{array}{r} 3\ 2\ 1 \\ -\ 3\ 9 \\ \hline \end{array}$$

8.
$$\begin{array}{r} 4\ 5\ 9 \\ -2\ 4\ 1 \\ \hline \end{array}$$

9.
$$\begin{array}{r} 6\ 1\ 4 \\ -1\ 9\ 6 \\ \hline \end{array}$$

10. Michael's book has 235 pages. He has read 167 pages. How many pages does he still have to read?

_____

11. Alexandria collected 300 pennies, while her brother collected 245 pennies. How many more pennies does she have than her brother has?

_____

12. While on vacation, Riley saw many tall buildings. One building was 425 feet tall, and another was 580 feet tall. What is the difference in the heights of the buildings?

_____

24B

Subtract using regrouping. Check by adding.

1. 716
    − 58

2. 668
    − 284

3. 163
    − 72

4. 641
    − 49

5. 372
    − 106

6. 987
    − 789

7. 705
    − 60

8. 500
    − 250

9. 471
    − 106

10. Laura had $275 to spend on clothes. If she buys a coat that costs $116, how much does she have left to spend?

_____

11. A squirrel collected 553 acorns and buried them last fall. He dug up and ate 378 of them during the winter. How many acorns are left in the ground?

_____

12. Grace Joy planned a 325-mile trip. The first day she traveled 185 miles. How far did she have left to travel?

_____

24C

Subtract using regrouping. Check by adding.

1.   583
     – 64

2.   203
     – 192

3.   200
     – 98

4.   319
     – 30

5.   631
     – 429

6.   419
     – 138

7.   103
     – 25

8.   572
     – 390

.9   333
     – 144

10. Eight hundred seventy-five people visited the new store. Only eighty people went home without buying something. How many people bought something in the new store?

_____

11. Joshua collected 200 baseball cards. He gave 115 of them to his best friend. How many baseball cards does Joshua have now?

_____

12. Isabella bought 375 yards of yarn for a knitting project. When she was finished knitting, she had 18 yards left over. How many yards of yarn did she use?

_____

Subtract using regrouping as needed. Check by adding.

1.
```
  1 0 0
 -  7 5
```

2.
```
  9 0 8
 - 2 9 1
```

3.
```
  2 5 6
 - 1 3 8
```

4.
```
  1 9 9
 -  8 2
```

5.
```
   6 3
 -  2 4
```

6.
```
   4 1
 -  3 5
```

Give the time with hours and minutes.

7.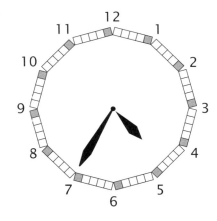

_____

8.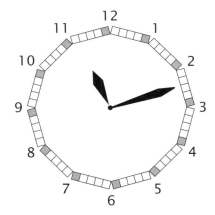

_____

Add.

9.　　 4 3 6
　　 + 1 2 2

10.　　 2 9 8
　　 + 5 3 9

11.　　 2, 9 9 9
　　 + 3, 1 1 1

12. Jayden caught 175 fireflies last night and put them in his room. This morning he counted only 98 fireflies. How many are missing?

_____

13. We drove 212 miles this morning and 362 miles this afternoon. How many miles did we drive today?

_____

14. Mom bought a chair for $215 and an end table for $134. How much did she have left over if she started with $400 to spend on furniture?

_____

24E

Subtract using regrouping as needed. Check by adding.

1.　 1 5 2
　　 − 8 7

2.　 3 2 0
　　 − 1 1 8

3.　 8 0 3
　　 − 3 0 0

4.　 1 9 3
　　 − 6 7

5.　 4 3
　　 − 1 7

6.　 6 4
　　 − 3 8

Give the time with hours and minutes.

7.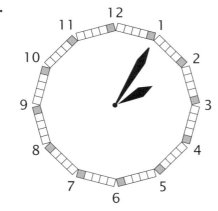

_____

8.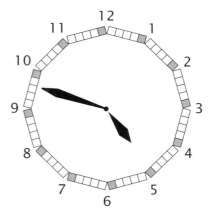

_____

Add.

9.   2 1 2
   + 3 6 2

10.   3 5 6
    + 4 8 1

11.   1, 2 7 6
    + 7, 3 9 1

12. Janna read a book with 321 pages, while Cassie read a book with 215 pages. What was the difference in the number of pages they read?

   _____

13. If Farmer John had 665 cows and 133 calves in the barn, how many animals did he have in the barn?

   _____

14. Mr. Herr sold 658 ice cream cones yesterday. He sold 314 chocolate cones and 219 vanilla cones. The rest were strawberry. How many strawberry-flavored ice cream cones did he sell?

   _____

24F

Subtract using regrouping as needed. Check by adding.

1.
```
  1 0 0
 −  4 7
```

2.
```
  1 4 1
 − 1 1 3
```

3.
```
  2 2 0
 − 1 6 4
```

4.
```
  1 5 0
 −  9 8
```

5.
```
   7 0
 − 2 0
```

6.
```
   3 2
 − 1 8
```

Give the time with hours and minutes.

7.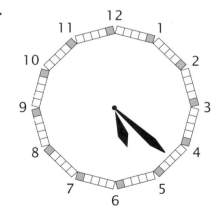

_____

8.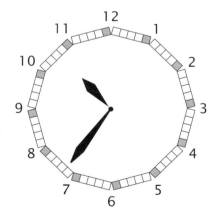

_____

Add.

9.
```
    4 0 0
  + 3 7 6
```

10.
```
    2 8 8
  + 1 3 7
```

11.
```
    2, 1 4 6
  + 1, 4 0 8
```

12. Five hundred eighty-one people went to the concert. Four hundred ninety-nine people loved the music, but the rest were unhappy. How many were unhappy?

_____

13. Robin spent $126 for groceries the first week and $132 the second week. How much did she spend in all for groceries?

_____

14. Abby collects little toy dogs. She already had 45 of them at the beginning of this year. She bought 11 more, got 5 for her birthday, and 8 for Christmas. How many more must she collect before she has 100 dogs?

_____

25A

Fill in the blanks.

1. The second month is _____ .

2. The _____ day is Monday.

Matching.

3. first   Tuesday

4. second   Sunday

5. third   Friday

6. fourth   Thursday

7. fifth   Wednesday

8. sixth   Saturday

9. seventh   Monday

How many days are in each month?

10. January _____   11. December _____

12. June _____

Tell how many.

13. |||| _____   14. ||| _____

15. ||||| ||| _____

Show the number with tally marks.

16. 19 _____

17. 30 _____

18. 29 _____

19. Aaron had the hiccups and kept track of them for an hour. Show his tally marks if he hiccuped 35 times.

_____

20. The first day of summer is in June. Which month is that?

_____

25B

Fill in the blanks.

1. The fourth day is _____ .

2. The _____ day is Saturday.

3. The tenth month is _____ .

Matching.

4. first	April
5. second	February
6. third	June
7. fourth	January
8. fifth	March
9. sixth	May

How many days are in each month?

10. February _____   11.   August _____

12. November _____

Tell how many.

13. ||||| ||||| | _____   14. ||||| | _____

15. ||||| ||||| ||||| _____

Show the number with tally marks.

16. 10 _____

17. 13 _____

18. 7 _____

19. Alex kept track of all the red cars he saw on the way to town. His tally looked like this: |||| |||| |||| ||.
How many red cars did he see?

_____

20. Which day is the first day of the week?

_____

Fill in the blanks.

1. The fourth month is _____ .

2. The _____ day is Tuesday.

3. The _____ day is Sunday.

Matching.

4. seventh   September

5. eighth   December

6. ninth   July

7. tenth   November

8. eleventh   October

9. twelfth   August

How many days are in each month?

10. April _____   11. March _____

12. July _____

Tell how many.

13. ||||| ||||| || _____   14. ||||| ||| _____

15. ||||| ||||| |||| _____

Show the number with tally marks.

16. 25 _____

17. 9 _____

18. 4 _____

19. Steve was born on the fifth day of the week. Which day is that?

_____

20. What holiday is on the twenty-fifth day of the twelfth month?

_____

Fill in the blanks.

1. The first month is _____ .

2. The _____ day is Wednesday.

3. The _____ day is Friday.

4. May has _____ days.

Add or subtract.

5.
```
   3 5 5
 -   2 6
 -------
```

6.
```
   6 1 8
 - 2 2 4
 -------
```

7.
```
   4 1 9
 - 3 5 7
 -------
```

8.
```
   3,6 2 9
 + 2,4 2 8
 ---------
```

Give the time with hours and minutes.

9.

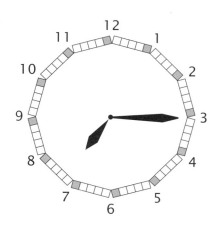

_____

10.

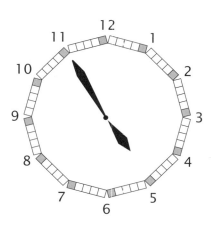

_____

11. Bill kept track of the number of times he had to erase mistakes on his math worksheet. Monday looked like this: ℍℍ ℍℍ ll; and Tuesday like this: ℍℍ lll. How many times did he have to erase in two days?

_____

12. Ryan had 5 red jelly beans, 17 green jelly beans, and 14 blue jelly beans. Find how many he had in all, and then write your answer with tally marks.

_____

13. Bria counted 131 days until the end of school. After 47 days had gone by, how many were still left?

_____

14. What is the perimeter of a square that is 11 feet on each side?

_____

Fill in the blanks.

1. The third month is _____ .

2. The sixth day is _____ .

3. The ninth month is _____ .

4. September has _____ days.

Add or subtract.

5.
```
  2 3 4
-   5 2
```

6.
```
  5 9 0
- 4 1 8
```

7.
```
  9 1 0
-   7 5
```

8.
```
  1,8 3 1
+ 5,4 3 9
```

Give the time with hours and minutes.

9.

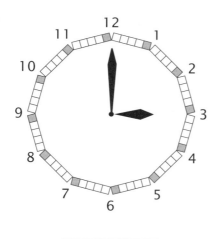

_____

10.

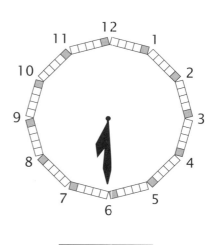

_____

11. Carla kept track of how many times it rained for a month: IIII IIII II. How many times did it rain?

_____

12. Use tally marks to record your age.

_____

13. Ian flew in an airplane three times last month. The first time he flew 3,451 miles, the second time he flew 2,163 miles, and the last time he flew 1,999 miles. How many miles did Ian fly in all?

_____

14. Sandi's birthday is on the first day of the eighth month. In what month is her birthday?

_____

Fill in the blanks.

1. The fifth month is _____ .

2. The eleventh month is _____ .

3. The _____ day is Friday.

4. October has _____ days.

Add or subtract.

5.
```
  5 0 7
-   4 1
```

6.
```
  1 8 2
- 1 4 6
```

7.
```
  9 0 0
-   2 5
```

8.
```
  7, 1 4 7
+ 8, 7 1 3
```

Give the time with hours and minutes.

9.

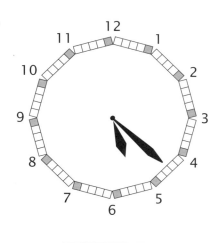

_____

10.

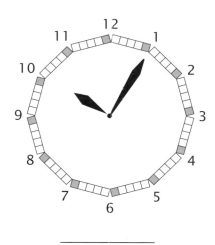

_____

11. Ben and David kept track of the books they read for a month. Ben: IIII IIII IIII IIII, and David: IIII IIII IIII IIII I. Who read more books?

_____

12. Our family enjoys Saturdays. What day of the week is that?

_____

13. Our house is 110 years old. Our neighbor's house is 35 years old. What is the difference in the ages of the two houses?

_____

14. The three sides of a triangle are 6 inches, 10 inches, and 13 inches. What is the perimeter of the triangle?

_____

26A

Subtract using regrouping. Check by adding in whichever way you like best. The first two are done for you.

1.
$$\begin{array}{r} 1,\overset{4}{\cancel{5}}\,\overset{1}{2}\,6 \\ -\quad 8\,7\,3 \\ \hline 6\,5\,3 \\ \text{\small mmmmm} \\ 1,5\,2\,6 \end{array}$$

check
$$\begin{array}{r} 8\,7\,3 \\ +\ \ 6\,5\,3 \\ \hline 1,\,5\,2\,6 \end{array}$$

2.
$$\begin{array}{r} \overset{3}{\cancel{4}},\overset{19}{\cancel{0}}\,\overset{19}{\cancel{0}}\,\overset{1}{0} \\ -\ 2,\,2\,5\,6 \\ \hline 1,\,7\,4\,4 \\ \text{\small mmmmm} \\ 4,\,0\,0\,0 \end{array}$$

check
$$\begin{array}{r} 2,\,2\,5\,6 \\ +\,1,\,7\,4\,4 \\ \hline 4,\,0\,0\,0 \end{array}$$

3.
$$\begin{array}{r} 5,3\,3\,3 \\ -\,1,1\,8\,6 \\ \hline \end{array}$$

4.
$$\begin{array}{r} 4,2\,8\,4 \\ -\quad 9\,5\,5 \\ \hline \end{array}$$

5.
$$\begin{array}{r} 8,2\,6\,3 \\ -\,3,1\,4\,9 \\ \hline \end{array}$$

6.
$$\begin{array}{r} 3,2\,6\,4 \\ -\,2,5\,8\,2 \\ \hline \end{array}$$

7.
$$\begin{array}{r} 7,2\,4\,1 \\ -\quad 3\,7\,8 \\ \hline \end{array}$$

8.
$$\begin{array}{r} 6,0\,0\,0 \\ -\,5,1\,3\,9 \\ \hline \end{array}$$

9.
$$\begin{array}{r} 6,7\,3\,2 \\ -\,3,1\,5\,2 \\ \hline \end{array}$$

If you have trouble keeping numbers in the right places, you can use notebook paper turned sideways.

10. Micah counted 1,465 ants. He watched 906 of them go back into their hole. How many ants are left for Micah to watch?

   _____

11. It is 2,375 miles to Grandmother's house. If Anah has already traveled 1,490 miles, how far does she still have to travel?

   _____

12. While on vacation, Kaylee saw one tree that was 1,760 years old and another tree that was 1,588 years old. What was the difference in their ages?

   _____

26B

Subtract using regrouping. Check by adding.

1.
$$
\begin{array}{r}
5,089 \\
-632 \\
\hline
\end{array}
$$

2.
$$
\begin{array}{r}
7,321 \\
-2,514 \\
\hline
\end{array}
$$

3.
$$
\begin{array}{r}
9,000 \\
-1,287 \\
\hline
\end{array}
$$

4.
$$
\begin{array}{r}
7,111 \\
-232 \\
\hline
\end{array}
$$

5.
$$
\begin{array}{r}
5,361 \\
-3,765 \\
\hline
\end{array}
$$

6.
$$
\begin{array}{r}
7,214 \\
-1,108 \\
\hline
\end{array}
$$

7.
$$
\begin{array}{r}
6,403 \\
-257 \\
\hline
\end{array}
$$

8.
$$
\begin{array}{r}
8,765 \\
-3,085 \\
\hline
\end{array}
$$

9.
$$
\begin{array}{r}
4,987 \\
-3,732 \\
\hline
\end{array}
$$

10. Mom saw a set of living room furniture that cost $1,579. She has $890 saved. How much more money does she need to buy the set?

_____

11. Jeremy and Jeffrey lived far apart and wanted to visit each other. Jeremy traveled 3,451 miles, and Jeffrey traveled 2,999 miles. How many more miles did Jeremy travel than Jeffrey?

_____

12. Joanne was born in 1976, and her grandfather was born in 1917. How old was her grandfather when Joanne was born? (Dates are a way of counting years. Subtract to find the number of years between two dates.)

_____

Subtract using regrouping. Check by adding.

1.
```
  1,6 5 0
 -  9 4 3
```

2.
```
  8,2 0 0
 -2,8 1 7
```

3.
```
  5,2 2 1
 -4,7 4 0
```

4.
```
  8,7 6 5
 -  6 7 8
```

5.
```
  2,9 0 6
 -1,0 8 8
```

6.
```
  6,4 2 9
 -3,5 8 7
```

7.
```
  3,4 0 5
 -  1 5 9
```

8.
```
  7,0 0 1
 -5,9 9 1
```

9.
```
  9,3 2 4
 -8,3 5 8
```

10. Kate learned that Columbus sailed to America in 1492. If it is 2004 now, how long ago did Columbus sail?

   _____

11. Kimberly had $3,600 in her savings account. If she took out $150, how much was left in her savings?

   _____

12. A farmer harvested 2,562 bushels of corn. He sold 1,600 bushels. How many bushels of corn does he have left?

   _____

Subtract using regrouping as needed. Check by adding.

1.
```
  4,3 0 5
 -   2 8 9
```

2.
```
  5,8 4 0
 -3,9 1 4
```

3.
```
  7,0 1 3
 -1,5 2 3
```

4.
```
  2 0 0
 -  3 4
```

5.
```
  7 7 2
 -4 1 6
```

6.
```
  2 9
 -1 5
```

Show the number with tally marks.

7. 8 _____

8. 31 _____

9. 14 _____

Give the time with hours and minutes.

10.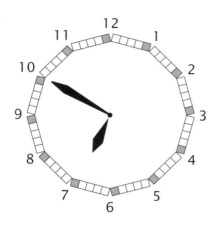

11.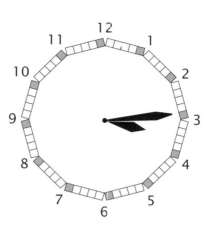

_____   _____

12. _____ is the third day of the week.

13. December is the _____ month of the year.

14. Kara is five feet tall. How many inches tall is she?

_____

15. Stephanie traveled 451 miles the first day and 385 miles the second day. How many miles must she still travel to finish her 1,000-mile trip?

_____

Subtract using regrouping as needed. Check by adding.

1.
$$1,800$$
$$-\ \ \ 176$$

2.
$$6,950$$
$$-4,867$$

3.
$$2,093$$
$$-2,075$$

4.
$$741$$
$$-\ 53$$

5.
$$500$$
$$-211$$

6.
$$48$$
$$-19$$

Show the number with tally marks.

7. 5 _____

8. 24 _____

9. 12 _____

Give the time with hours and minutes.

10.

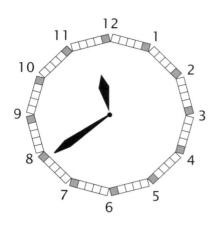

_____

11.

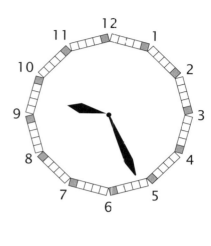

_____

12. _____ is the tenth month of the year.

13. Saturday is the _____ day of the week.

14. Alison collected 673 acorns. Chucky collected 415 acorns and added them to the pile. A few days later they could find only 223 acorns. How many do you think the squirrels took away?

_____

15. What is the perimeter of a square that is 10 inches on each side?

_____

26F

Subtract using regrouping as needed. Check by adding.

1.
```
  8,9|0|1
-   |7|9|2
```

2.
```
  9,0|1|2
- 8,9|9|9
```

3.
```
  1,2|3|4
- 1,0|4|5
```

4.
```
  6 7 0
-   6 9
```

5.
```
  7 6 5
- 5 7 8
```

6.
```
    3 2
-   2 5
```

Show the number with tally marks.

7. 11 _____

8. 33 _____

9. 26 _____

Give the time with hours and minutes.

10.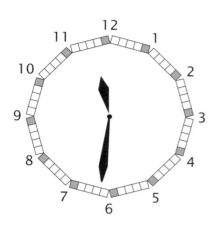

11.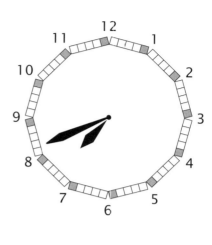

_____   _____

12. _____ is the fifth day of the week.

13. June is the _____ month of the year.

14. Christel called two friends on Monday, four on Tuesday, five on Wednesday, six on Thursday, and eight on Friday. How many phone calls did she make on those five days?

_____

15. What is the perimeter of a rectangle that is six inches long and three inches high?

_____

Subtract the money. The first one is done for you.

1. $1.$^{2}3^{1}4$
    − . 7 5
    $. 5 9

2. $9. 0 0
    − 2. 6 7

3. $ .4 6
    − .1 0

4. $6. 1 4
    − 1. 2 1

5. $2. 4 5
    − 1. 3 8

6. $ .8 0
    − .2 4

7. $5. 1 9
    − 3. 7 2

8. $ 2 6. 0 0
    − 1 1. 9 2

9. $ 3 4. 4 7
    − 2 2. 0 9

10. Ethan had $7.15 when he went into the store and $2.98 when he came out. How much did he spend?

   _____

11. Bria had $9.46 in her purse. She decided to spend $4.91 on a gift for Duncan. How much did she have left?

   _____

12. My boots cost $45.50 and my shoes cost $34.99. How much more did my boots cost?

   _____

27B

Subtract the money.

1.　　$2.65
　　－　.38

2.　　$7.25
　　－　1.89

3.　　$ .10
　　－.08

4.　　$1.00
　　－.77

5.　　$6.19
　　－　3.58

6.　　$ .93
　　－.45

7.　　$2.47
　　－　1.09

8.　　$35.60
　　－　21.90

9.　　$74.82
　　－　36.25

10. Matthew and Josh bought CDs. Matthew spent $25.69 and Josh spent $31.16. How much more did Josh spend?

_____

11. The game that Abby wants to buy costs $39.67. She has saved $20.75. How much more does she need to save?

_____

12. Caitlyn earned $13.05 helping her mom. After she went shopping, she had only $.36 left. How much did Caitlyn spend?

_____

Subtract the money.

1. $5.13
    − .48

2. $8.15
    − 7.09

3. $.73
    − .17

4. $3.00
    − .81

5. $9.32
    − 6.14

6. $.86
    − .55

7. $4.75
    − 2.06

8. $20.15
    − 13.21

9. $98.17
    − 25.18

10. Sue took $10.00 to the store to buy cloth for a quilt. If she came home with $1.25, how much did she spend?

_____

11. Terri wants to buy a gift for Mindy that costs $53.95. If she has saved $45.00 already, how much more does she need?

_____

12. Bananas cost 45¢ a pound and apples cost 61¢ a pound. How much more would it cost to buy a pound of apples?

_____

27D

Subtract.

1. $8.91
  − .76

2. $6.82
  − 1.61

3. 7,712
  − 5,872

Add.

4. 184
  + 68

5. 255
  + 177

6. 3,011
  + 1,895

Matching.

7. first   Wednesday

8. second   Friday

9. third   Saturday

10. fourth   Sunday

11. fifth   Thursday

12. sixth   Monday

13. seventh   Tuesday

Skip count by two.

14. 2, _____, _____, _____, _____, _____, _____, _____, _____, _____

15. Rick kept track of the birds that came to his feeder one morning: ⊞ ⊞ ⊞ II.
How many birds did he see?

_____

16. Chucky got $5.00 from each of his four sisters for his birthday. Use column addition to find out how much he got in all. If Chucky spent $6.95 for a ball, how much money did he have left?

_____

Subtract.

1. $ 4 . 4 2
      − .3 9

2. $ 1 . 2 0
      − 1 . 1 8

3.   4,5 0 3
      − 2,9 0 1

Add.

4.   8 3 6
     + 1 7

5. $ 3 . 4 5
     + 7 . 1 8

6.   1,8 1 9
     + 4,4 2 8

Matching.

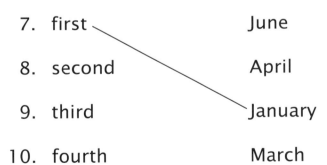

7.   first   June

8.   second   April

9.   third   January

10.   fourth   March

11.   fifth   February

12.   sixth   May

Skip count by five.

13. 5, _____, _____, _____, _____, _____, _____, _____, _____, _____

14. Hannah kept track of the days until the package she had ordered came in the mail: ‖‖ ‖‖ ‖‖ ‖‖ ‖‖ ‖‖.

How many days did she wait?

_____

15. Drew did two chores. He earned $5.00 and $6.50. Cameron did one chore. He earned $10.20. Find out which boy earned more. Can you remember how to use an inequality to show your answer?

_____

Subtract.

1. $ 3 . 1 4
    − 1 . 0 9

2. $ 2 . 6 2
    − 1 . 3 8

3. $ 1 9 . 0 7
    − 1 5 . 2 1

Add.

4. 9 7 7
    + 1 3

5. $ 6 . 5 4
    + 9 . 5 4

6. 1 , 7 6 4
    1 , 9 1 3
    + 2 , 3 8 4

Matching.

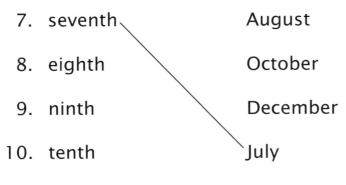

7. seventh   August

8. eighth   October

9. ninth   December

10. tenth   July

11. eleventh   November

12. twelfth   September

Skip count by ten.

13. 10, _____, _____, _____, _____, _____, _____, _____, _____, _____

14. Use tally marks to show the number of days in August.

   _____

15. A fence went all around the perimeter of a triangle that had sides of 35 feet, 25 feet, and 19 feet. Ten feet of fence got knocked down when Bill hit it with his go-cart. How many feet of fence are left standing?

   _____

Subtract using regrouping. Check your work using your favorite way. The first one is done for you.

1.
$$\begin{array}{r} {}^{1}\cancel{2}\,{}^{1}1,{}^{6}\cancel{7}\,{}^{1\!5}\cancel{6}\,{}^{1}2 \\ -\ 1\ 4,\ 3\ 7\ 5 \\ \hline 7,\ 3\ 8\ 7 \end{array}$$
$$\begin{array}{r} 2\ 1,\ 7\ 6\ 2 \end{array}$$

check
$$\begin{array}{r} 1\ 4,\ 3\ 7\ 5 \\ +\ \ \ 7,\ 3\ 8\ 7 \\ \hline 2\ 1,\ 7\ 6\ 2 \end{array}$$

2.
$$\begin{array}{r} 8\,4,5\,2\,8 \\ -\ 6\,4,0\,2\,5 \\ \hline \end{array}$$

3.
$$\begin{array}{r} 3\,5,1\,9\,4 \\ -\ 3\,1,4\,8\,6 \\ \hline \end{array}$$

4.
$$\begin{array}{r} 4\,2,3\,5\,5 \\ -\ 2\,1,4\,7\,2 \\ \hline \end{array}$$

5.
$$\begin{array}{r} 7\,1,6\,2\,1 \\ -\ 4\,1,5\,7\,3 \\ \hline \end{array}$$

6.
$$\begin{array}{r} 5\,6,4\,0\,8 \\ -\ 2\,4,3\,7\,9 \\ \hline \end{array}$$

7. 45,900 people used to live in our town, but 22,175 people moved away. How many people live in the town now?

_____

8. 31,231 tadpoles hatched out of their eggs, but 19,452 were eaten before they could grow up to be frogs. How many tadpoles were left to become frogs?

_____

9. The distance around the earth is about 25,000 miles. Thomas set out to fly all the way around in his plane. If he has flown 19,000 miles so far, how many miles does he still have to go?

_____

10. Ruth thought she would earn $45,575 last year. Since she was sick a lot and couldn't go to work, she earned only $38,196. What is the difference in the two amounts?

_____

28B

Subtract using regrouping. Check your work.

1.
$$
\begin{array}{r}
6\,8{,}5\,1\,1 \\
-\ 1\,9{,}3\,3\,3 \\
\hline
\end{array}
$$

2.
$$
\begin{array}{r}
2\,5{,}3\,8\,2 \\
-\ 1\,1{,}2\,5\,5 \\
\hline
\end{array}
$$

3.
$$
\begin{array}{r}
4\,7{,}4\,6\,0 \\
-\ 3\,3{,}4\,1\,9 \\
\hline
\end{array}
$$

4.
$$
\begin{array}{r}
7\,0{,}0\,0\,0 \\
-\ 1\,9{,}9\,9\,9 \\
\hline
\end{array}
$$

5.
$$
\begin{array}{r}
6\,5{,}2\,4\,2 \\
-\ 2\,1{,}1\,3\,5 \\
\hline
\end{array}
$$

6.
$$
\begin{array}{r}
5\,4{,}9\,6\,7 \\
-\ 4\,2{,}7\,1\,8 \\
\hline
\end{array}
$$

7. 10,752 people voted in the town election. Of that number, 6,834 people voted for Mr. Jones for mayor. How many people voted for someone else?

_____

8. A farmer planted 55,780 seeds of corn. Because it was so dry, 29,592 seeds did not sprout. How many corn plants did the farmer get?

_____

9. Rebecca figured out that her one-mile walk was 63,360 inches long. Clara walked only 31,680 inches. What is the difference in the number of inches they walked?

_____

10. Ben is thinking of buying a new car. One car he likes costs $16,899, and another costs $22,980. What is the difference in price between the two cars?

_____

28C

Subtract using regrouping. Check your work.

1.
```
  1 3 , 5 2 2
- 1 2 , 0 4 8
```

2.
```
  7 9 , 1 4 7
- 2 4 , 3 1 2
```

3.
```
  5 3 , 6 4 1
- 2 0 , 4 6 5
```

4.
```
  8 4 , 6 7 0
- 3 5 , 9 0 1
```

5.
```
  1 3 , 5 0 6
-   9 , 9 5 1
```

6.
```
  4 1 , 0 0 3
- 1 0 , 5 2 3
```

7. The odometer in a car measures how many miles the car has been driven. When Elizabeth left on vacation, the odometer in her car read 19,761 miles. When she got home, it read 23,295 miles. How many miles had she driven?

_____

8. The plastics company made 53,600 green unit blocks for Math-U-See. If 48,096 have been packed in the boxes to sell, how many unit blocks are left to pack?

_____

9. A scientist estimated that 60,000 fish lived in a lake. He thought that about 22,000 of the fish were trout. If he is correct, how many fish are not trout?

_____

10. Katherine wants to buy a house that costs $95,899. She has $26,900. How much more money does she need?

_____

28D

Subtract using regrouping as needed. Check your work.

1.
$$\begin{array}{r} 9\,6,0\,2\,1 \\ -\ 4\,5,6\,3\,5 \\ \hline \end{array}$$

2.
$$\begin{array}{r} 8\,0,0\,0\,0 \\ -\ 7\,9,9\,9\,8 \\ \hline \end{array}$$

3.
$$\begin{array}{r} \$\,5\,0.7\,5 \\ -\ \ 1\,0.2\,5 \\ \hline \end{array}$$

4.
$$\begin{array}{r} \$\,3\,9.1\,3 \\ -\ \ 2\,2.4\,6 \\ \hline \end{array}$$

5.
$$\begin{array}{r} \$\,1\,9.1\,5 \\ -\ \ \ 7.9\,9 \\ \hline \end{array}$$

Fill in the oval with >, <, or =.

6. 3 $\bigcirc$ 4

7. 2 + 5 $\bigcirc$ 2 + 4

8. 3 + 6 $\bigcirc$ 6 + 3

Add.

9.
```
   1 2 5
   2 8 5
     4 4
 + 1 6 1
```

10.
```
   3 1 5
   2 0 0
   3 9 4
 + 1 5 5
```

11.
```
   4 4 9
   1 3 1
   5 0 3
 +   6 7
```

12. _____ is the first day of the week.

13. March is the _____ month of the year.

14. In April, Mom went grocery shopping four times. She spent $123.45, $210.13, $75.21, and $103.82. How much did Mom spend for food during that month?

   _____

15. A shepherd had 100 sheep, but when he counted them one night he found only 99 sheep. How many sheep were lost?

   _____

Subtract using regrouping as needed. Check your work.

1.
$$
\begin{array}{r}
1\,1{,}4\,3\,5 \\
-\ 1\,0{,}6\,8\,2 \\
\hline
\end{array}
$$

2.
$$
\begin{array}{r}
4\,0{,}7\,3\,4 \\
-\ 2\,7{,}1\,5\,6 \\
\hline
\end{array}
$$

3.
$$
\begin{array}{r}
\$\,2\,1.\,1\,1 \\
-\ \ 1\,9.\,8\,9 \\
\hline
\end{array}
$$

4.
$$
\begin{array}{r}
\$\,7\,8.\,0\,4 \\
-\ 3\,5.\,4\,0 \\
\hline
\end{array}
$$

5.
$$
\begin{array}{r}
\$\,6\,2.\,0\,0 \\
-\ 5\,1.\,1\,9 \\
\hline
\end{array}
$$

Fill in the oval with >, <, or =.

6. 14 $\bigcirc$ 7 + 7

7. 9 – 7 $\bigcirc$ 7 – 4

8. 63 $\bigcirc$ 36

Add.

9.
$$
\begin{array}{r}
7\,0\,4 \\
3\,0\,0 \\
2\,6 \\
+\quad\ \ 9 \\
\hline
\end{array}
$$

10.
$$
\begin{array}{r}
2\,4\,1 \\
1\,1\,9 \\
5\,6\,2 \\
+\ 5\,0\,8 \\
\hline
\end{array}
$$

11.
$$
\begin{array}{r}
3\,2\,5 \\
7\,6\,3 \\
1\,2\,5 \\
+\quad 4\,6 \\
\hline
\end{array}
$$

12. _____ is the eighth month of the year.

13. Friday is the _____ day of the week.

14. 10,456 spiderlings hatched in my garden. 8,912 of them got eaten by the birds. How many grew up to be big spiders?

_____

15. Farmer Brown has 25 cows, 89 sheep, and 5 horses. How many animals does he have in all?

_____

Subtract using regrouping as needed. Check your work.

1.
$$\begin{array}{r} 7\,9,3\,0\,4 \\ -\ 6\,1,0\,8\,2 \\ \hline \end{array}$$

2.
$$\begin{array}{r} 5\,5,0\,0\,0 \\ -\ 4\,8,1\,2\,3 \\ \hline \end{array}$$

3.
$$\begin{array}{r} \$\,1\,7.9\,9 \\ -\ \ \ 5.1\,5 \\ \hline \end{array}$$

4.
$$\begin{array}{r} \$\,4\,3.6\,0 \\ -\ 1\,3.7\,9 \\ \hline \end{array}$$

5.
$$\begin{array}{r} \$\,2\,8.7\,4 \\ -\ 1\,3.2\,5 \\ \hline \end{array}$$

Fill in the oval with >, <, or =.

6. 42 $\bigcirc$ 78

7. 14 + 2 $\bigcirc$ 13 − 2

8. 2 + 1 $\bigcirc$ 1 + 2

Add.

9.
```
  9 1 0
  2 6 2
  3 6 6
+ 1 4 8
```

10.
```
  3 5 5
  1 7 6
  7 5 5
+   2 3
```

11.
```
  4 3 1
  5 2 9
  6 1 1
+ 5 7 6
```

12. _____ is the fourth day of the week.

13. May is the _____ month of the year.

14. Caleb got $34.00 for his birthday and $25.00 for Christmas. He spent $29.98 on a gift for his sister. How much money does he have left?

   _____

15. Joel counted cars during his ride. He counted 25 red cars, 17 black cars, 30 blue cars, and 8 white cars. How many cars did he count altogether?

   _____

29A

Fill in the missing numbers, and then write the number of gallons on the line. The first one is done for you.

1.

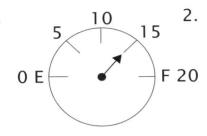

____15____ gallons

2.

2.

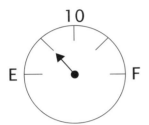

_____ gallons

3.

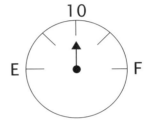

_____ gallons

How fast am I going on each speedometer?

4.

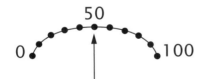

_____ mph   (The letters mph stand for miles per hour.)

5.

_____ mph

6.

_____ mph

Fill in the missing numbers, and then write the temperature on the line below. The first one is done for you.

7.

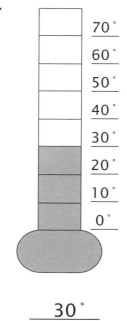

70°
60°
50°
40°
30°
20°
10°
0°

___30°___

8.

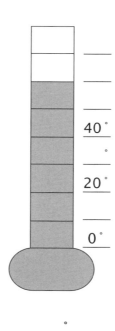

___
___
40°
°
20°
0°

_____ °

9.

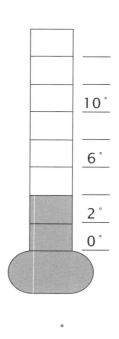

___
10°
6°
2°
0°

_____ °

29B

Fill in the missing numbers on the gauge, and then write the temperature on the line. The first one is done for you.

1.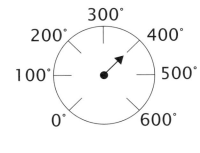

_____400_____°

2.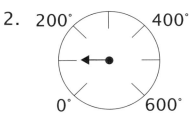

_____°

3.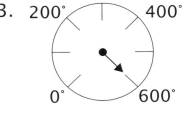

_____°

How fast am I going on each speedometer?

4.

_____ mph

5.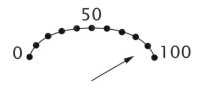

_____ mph

6.

_____ mph

Fill in the missing numbers, and then write the temperature on the line below.

7.

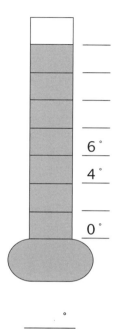

6°
4°
0°

_____°

8.

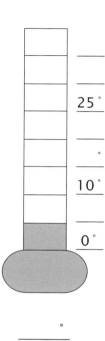

25°
°
10°
0°

_____°

9.

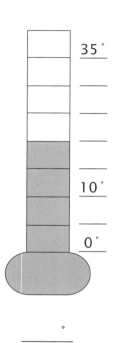

35°
10°
0°

_____°

Fill in the missing numbers, and then write the number of gallons on the line.

1.
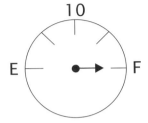

_____ gallons

2.
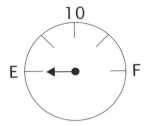

_____ gallons

3.
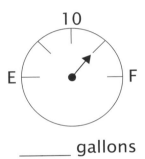

_____ gallons

How fast am I going on each speedometer?

4.
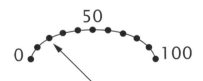

_____ mph

5.
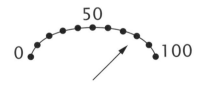

_____ mph

6.

_____ mph

Fill in the missing numbers, and then write the temperature on the line below.

7.

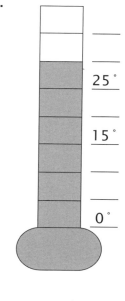

25°

15°

0°

_____ °

8.

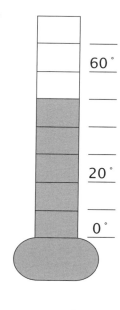

60°

20°

0°

_____ °

9.

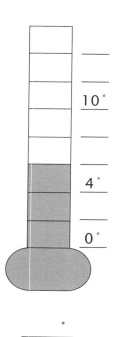

10°

4°

0°

_____ °

29D

Read the gauge or thermometer and write your answer on the line.

1.

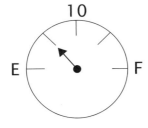

_____ gallons

2.

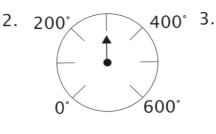

_____ °

3.

_____ °

4.

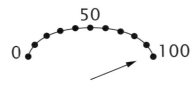

_____ mph

5.

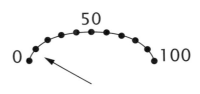

_____ mph

Add or subtract.

6.
$$
\begin{array}{r}
3\,4,1\,2\,7 \\
-\ 2\,5,1\,7\,9 \\
\hline
\end{array}
$$

7.
$$
\begin{array}{r}
5,6\,8\,0 \\
-\ 4,4\,5\,2 \\
\hline
\end{array}
$$

8.
$$
\begin{array}{r}
7,5\,6\,3 \\
+\ 5,1\,4\,8 \\
\hline
\end{array}
$$

Give the time with hours and minutes.

9.

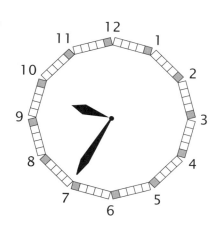

_____

10.

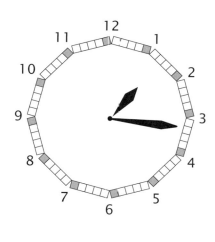

_____

11. Peter was born in 1962. How many years old was he in 2003?

_____

12. Use tally marks to show Peter's age (see #11).

_____

29E

Read the gauge or thermometer and write your answer on the line.

1.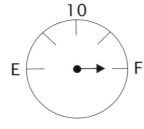

_____ gallons

2.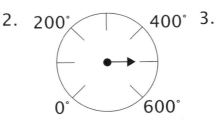

_____ °

3.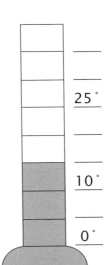

_____ °

4.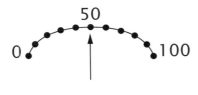

_____ mph

5.

_____ mph

Add or subtract.

6.
```
  1 7,5 3 9
- 1 5,6 5 9
```

7.
```
   7,0 5 8
 - 3,1 7 2
```

8.
```
   4,2 6 4
 + 2,7 9 1
```

QUICK TIP

Sequencing can be used to find your room in a hotel. The first digit in the room number tells which floor you are on. Once you are on the correct floor, the rooms are numbered in order.

Example: When you get off the elevator on the first floor, you see this sign.

Do you turn to the right or the left to find room 172?

101 – 145	146 – 190
←	→

Answer: You would turn right because 172 comes between 146 and 190.

9. Using the sign above, tell if you should turn right or left to find room 116. _____

10. Our country was 200 years old on July 4, 1976. What year did our country begin? _____

July is the _____ month.

29F

Read the gauge or thermometer and write your answer on the line.

1.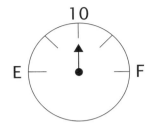

_____ gallons

2.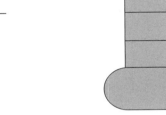

_____ °

3.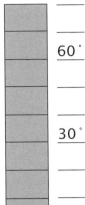

4.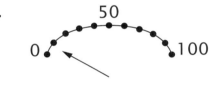

_____ mph

5.

_____ mph

Add or subtract.

6.
$$
\begin{array}{r}
9\,3,1\,0\,5 \\
-\ 6\,0,1\,2\,7 \\
\hline
\end{array}
$$

7.
$$
\begin{array}{r}
8,0\,0\,0 \\
-\ 2,1\,1\,1 \\
\hline
\end{array}
$$

8.
$$
\begin{array}{r}
6,4\,5\,6 \\
2,4\,3\,4 \\
+\ 1,8\,4\,9 \\
\hline
\end{array}
$$

Give the time with hours and minutes.

9.

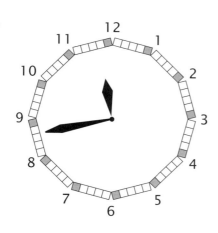

_____

10.

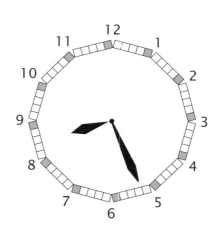

_____

11. Using the sign, tell if you should turn

right or left to find room 430.

401 – 450	451 – 490
←	→

_____

12. If the United States was 200 years old in 1876, how old is it now? (two steps)

_____

Use the graph to answer the questions. Follow the directions to make a line graph.

Snowfall Last Winter

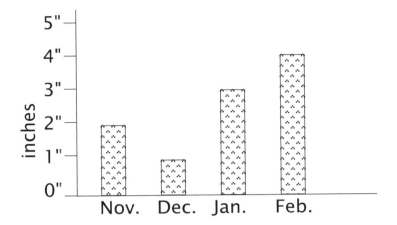

1. Which month had the most snow? _____

2. Which month had the least snow? _____

3. How much snow fell in November? _____

4. Which month had 3 inches of snow? _____

5. Make a line graph showing 3" in November, 2" in December, 4" in January, and 1" in February.

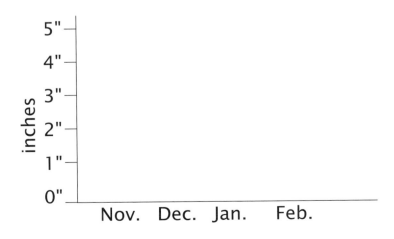

Line and bar graphs can be turned on their sides.

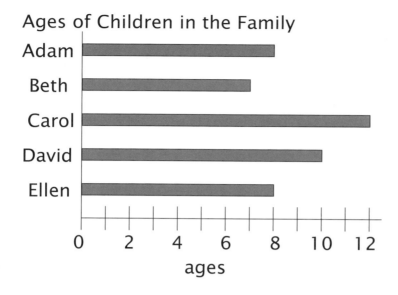

Ages of Children in the Family

6. Who is the oldest child? _____

7. How old is the youngest child? _____

8. How old is David? _____

9. Which two children are twins? _____

10. Make a line graph showing how old the children were two years ago.

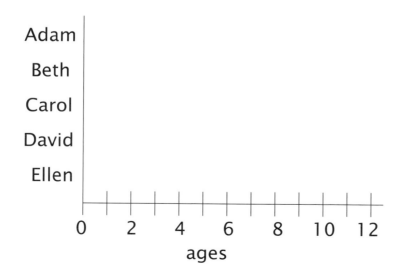

Use the graph to answer the questions. Follow the directions to make another graph.

Donut Sales

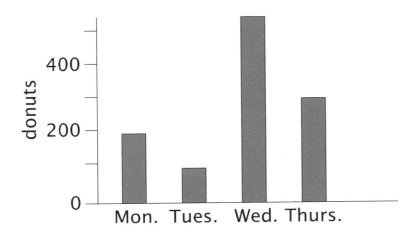

1. Which day had the most sales? _____

2. Which day had the fewest sales? _____

3. How many donuts were sold on Monday? _____

4. How many donuts were sold on Thursday? _____

5. The next week's sales were 200, 300, 100, and 400. Make a line graph.

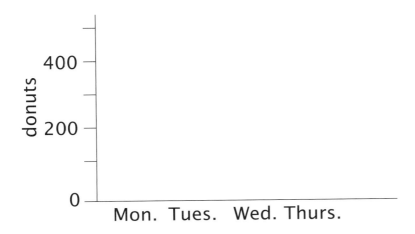

Colors of Cars Passing Bill's House

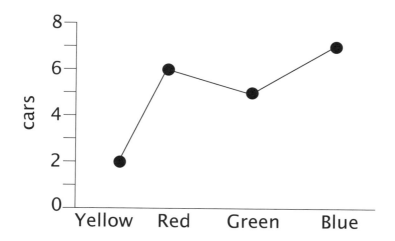

6. Which color did he see the most? _____

7. How many red cars? _____

8. How many green cars? _____

9. How many yellow cars? _____

10. The next day he saw 1 yellow car, 4 red cars, 8 green cars, and 5 blue cars. Make a bar graph.

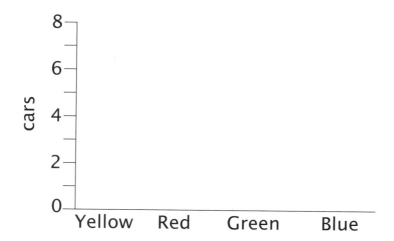

30C

Use the graph to answer the questions. Follow the directions to make another graph.

Miles Shane Drove Last Week

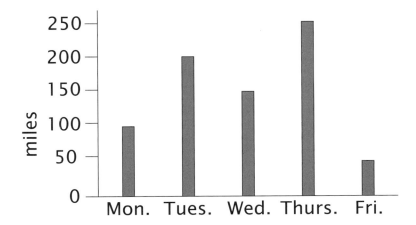

1. How many miles did he drive on Monday? _____

2. How many miles did he drive on Friday? _____

3. How many miles did Shane drive all week? _____

4. The next week Shane drove the following numbers of miles each day: 50, 150, 200, 100, and 250. Make a line graph.

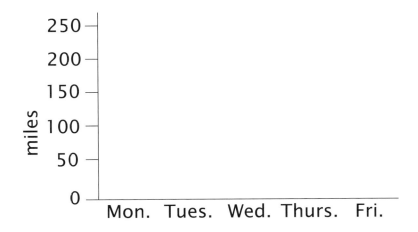

Books Sarah Read Each Month

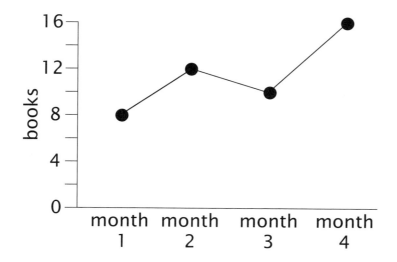

5. How many books did Sarah read the fourth month? _____

6. How many books did she read the second month? _____

7. How many books did Sarah read in four months? _____

8. The next four months Sarah read the following numbers of books each month: 10, 12, 8, and 12. Make a bar graph.

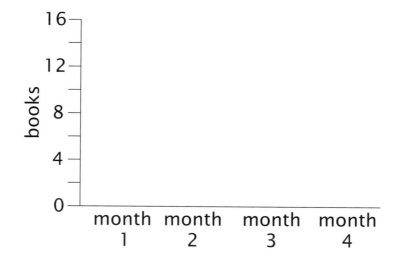

Use the graph to answer the questions. Follow the directions to make a line graph.

Isaac's Exercise Schedule

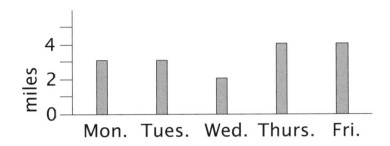

1. Which days did Isaac walk the farthest?
   _____ and _____

2. Which day did he walk the fewest miles? _____

3. How many miles did Isaac walk all week? _____

4. The same week Ethan walked the following numbers of miles each day: 3, 4, 1, 2, and 4. Make a line graph.

Ethan's Exercise Schedule

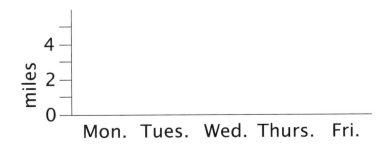

Read the gauge and write your answer on the line.

5.

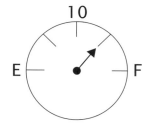

_____ gallons

6.

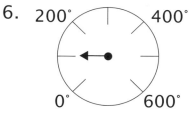

_____ °

Add or subtract.

7. $\begin{array}{r} \$\,7\,8.9\,0 \\ -\ 2\,1.9\,5 \\ \hline \end{array}$

8. $\begin{array}{r} \$\,1\,3.5\,4 \\ +\ \ \ 8.6\,3 \\ \hline \end{array}$

9. $\begin{array}{r} 5\,3,4\,1\,1 \\ -\ 1\,6,4\,2\,2 \\ \hline \end{array}$

10. Using the sign, tell if you should turn right or left to find room 761.

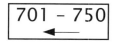

_____

11. Write the number shown by the tally marks: ⅢⅢ ⅢⅢ ⅢⅢ ⅢⅢ ⅢⅢ I.

_____

12. _____ is the third month.

Use the graph to answer the questions. Follow the directions to make a bar graph.

Monthly Rainfall

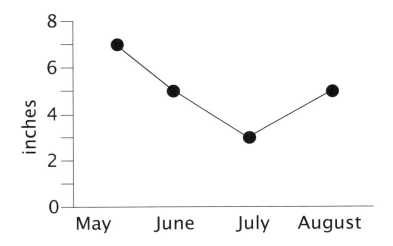

1. Which month had the most rain? _____

2. Which month had the least rain? _____

3. How much rain was there in August? _____

4. The next four months had the following amounts of rain: 4", 6", 8", and 5". Make a bar graph.

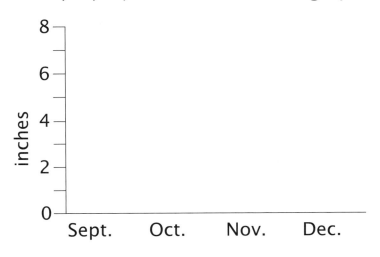

Read the gauge and write your answer on the line.

5.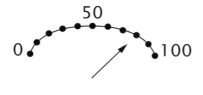

_____ mph

6.

_____ mph

Fill in the oval with >, <, or =.

7. 135 ◯ 531

8. 7 + 8 ◯ 5 + 5 + 5

9. 16 − 8 ◯ 5 + 4

10. Each side of a square measures 45 feet. What is the perimeter of the square?

_____

11. Jeff was born in 1945 and Jerry was born in 1972. What is the difference in their ages?

_____

12. Lisa drove 65 miles on Monday, 348 miles on Tuesday, and 179 miles on Wednesday. How many miles did she drive in those three days?

_____

Use the graph to answer the questions. Bar graphs can be turned on their sides.

Home Runs in April

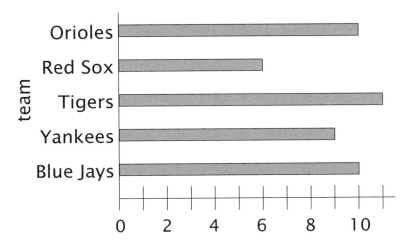

1. Which team hit the most home runs? _____

2. Which team hit the fewest home runs? _____

3. How many home runs did the Yankees hit? _____

4. Which two teams had the same number of home runs?

   _____ and _____

Read the gauge or thermometer and write your answer on the line below.

5.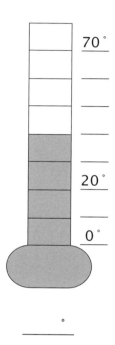

6.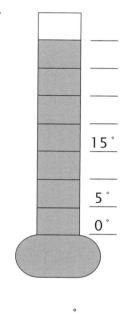

7.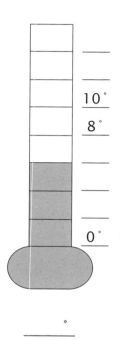

_____ °   _____ °   _____ °

Add or subtract.

8.
```
  7 6 , 5 4 1
- 3 2 , 1 5 3
```

9.
```
  6 , 0 0 0
- 3 , 2 6 5
```

10.
```
  3 , 1 2 8
  5 , 5 9 2
+ 4 , 5 1 7
```

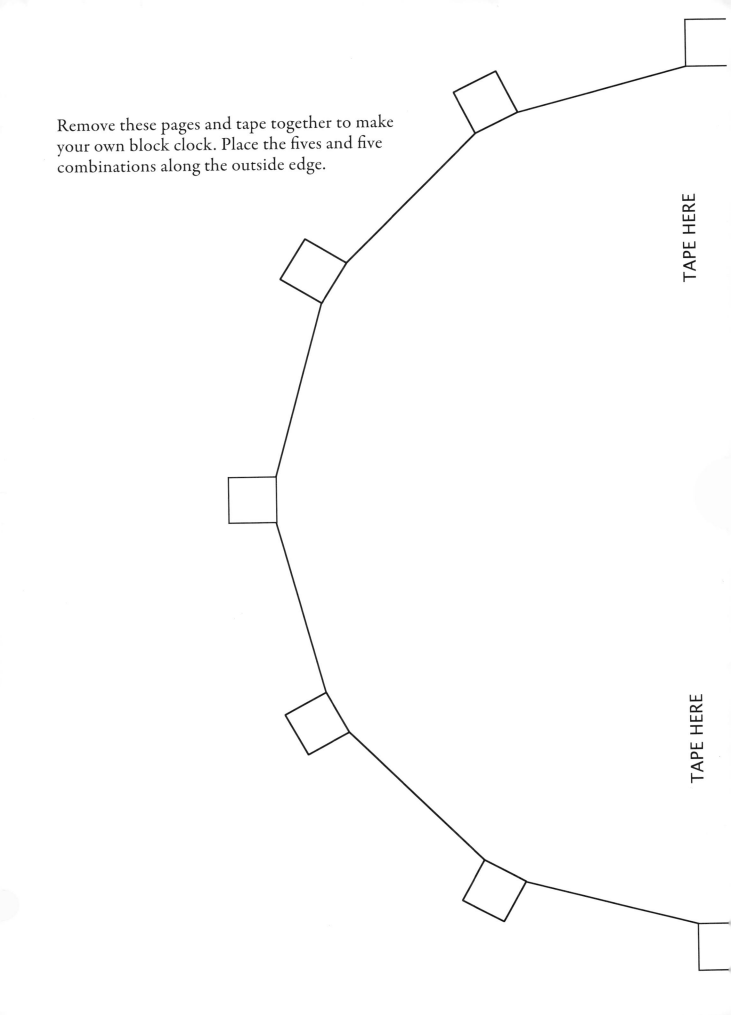

Remove these pages and tape together to make your own block clock. Place the fives and five combinations along the outside edge.

TAPE HERE

TAPE HERE

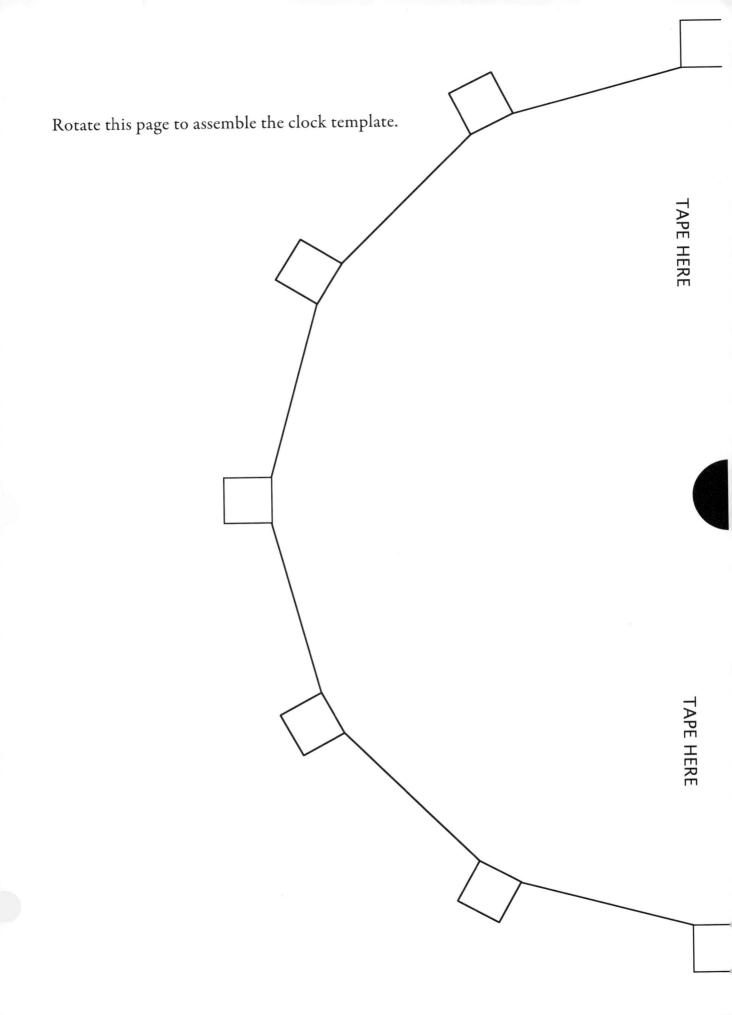

Rotate this page to assemble the clock template.

TAPE HERE

TAPE HERE

CONGRATULATIONS!!

To_____

On this_____ day of _____, 20___

You have just finished

Beta

You are becoming a math whiz!

Have fun doing the next book, which is

Gamma

Psssst. Don't forget to thank your Teacher!